MANAGING
CONFINED SPACES
IN
CONSTRUCTION

JAMES ROGERS

Library of Congress Control Number: 2017900669

CreateSpace Independent Publishing Platform, North Charleston, SC

Copyright © 2017 James D. Rogers

ISBN: 1523837209
ISBN-13: 978-1523837205

DISCLAIMER

This document is intended for the use of professionals competent to evaluate the significance and limitations of its contents and who will accept responsibility for the application of the materials it contains. The information contained herein is presented to educate the construction management professional about the dangers associated with confined spaces in construction and is meant to aid in the development of an appropriate system to manage confined spaces on the construction site. Proper evaluation of a confined space must be done by a Competent Person whose knowledge goes far beyond that which is contained in this publication, and entry into a permit-required confined space always requires the application of a space-specific entry plan developed by the Competent Person after a proper and thorough evaluation.

The Author, in publishing this document makes no warranty regarding the recommendations contained herein, including warranties of quality, workmanship or safety, express or implied, further including, but not limited to, implied warranties or merchantability and fitness for a particular purpose. THE AUTHOR SHALL NOT BE LIABLE FOR ANY DAMAGES, INCLUDING CONSEQUENTIAL DAMAGES BEYOND REFUND OF THE PURCHASE PRICE OF THIS PUBLICATION.

CONTENTS

PREFACE

There are many publications available related to confined spaces, however at the time of this publication, this author could find very few that deal with the management of confined spaces in the construction industry. Even the ones that do address construction tend to do so in the context of tank and silo construction or retrofits at existing facilities where spaces are well defined and controlled.

This book was designed to address the confined spaces that can be found on many typical construction sites. The ones that probably haven't been identified or classified yet because they didn't exist when the project started. The ones that get constructed without consideration of what work remains that needs to be done inside. The ones that many people in the construction industry do not recognize as confined spaces, such as manholes and catch basins. This book is about confined spaces in the construction industry.

1 INTRODUCTION

In May 2015, the United States Department of Labor's Occupational Safety and Health Administration (OSHA) issued new rules for the Construction Industry that changed the way confined spaces are managed on construction sites. Previously, a comprehensive set of rules existed that applied to General Industry, found in 29 CFR 1910.146, but that set of regulations specifically excluded construction work. Although many safety professionals would publicly state that these General Industry rules needed to be followed when entering spaces during construction that met the definition of a permit-required confined space, a formal Permit entry system is rarely seen on construction sites. In fact, construction companies could legitimately argue that there was "no such thing" as a permit-required confined space in construction because that term appeared nowhere in OSHA's Construction Industry regulations (29 CFR 1926), and the opening statement in the General Industry regulations (§1910.146) clearly states that it does not apply to any work covered in OSHA's 29 CFR 1926 Construction Industry regulations.

Of course, none of these arguments obviated an employer's legal responsibility under the "General Duty Clause" (29 USC 654) contained in the Occupational Safety and Health Act of 1970 (also known as Public Law 91-596) which states...

Each employer shall (1) furnish to each of his employees employment and a place of employment which are free from recognized hazards that are causing or are likely to cause death or serious physical harm to his employees; (2) shall comply with occupational safety and health standards promulgated under this Act.

However, in practice many construction employers have done little more than comply with the only other reference that was previously contained in the Construction Industry Standards (§1926.21(b)(6)(i)), which simply stated…

All employees required to enter into confined or enclosed spaces shall be instructed as to the nature of the hazards involved, the necessary precautions to be taken, and in the use of protective and emergency equipment required. The employer shall comply with any specific regulations that apply to work in dangerous or potentially dangerous areas.

Furthermore, the lack of a specific standard or set of rules for confined space entry in construction was sometimes used to defend a contractor's lack of written entry procedures when sending people into what should have been recognized as a permit-required confined space to perform repair or maintenance operations by simply categorizing the work as "construction".

With the issuance of the Construction Industry Standard known as 29 CFR 1926 Subpart AA, OSHA has closed that loophole and made the characterization of the work as General Industry work versus Construction Industry work a moot point. Both sets of industry Standards now contain definitions for confined spaces and permit-required confined spaces, and both Standards now contain a specific set of rules and procedures that must be followed prior to, during and after entry into hazardous confined spaces. In fact, OSHA has even stated in their compliance directives that contractors performing "construction" work at an existing facility that operates under the "general industry" regulations will be considered to be in compliance if they are following the new Construction Industry Standards contained in §1926 Subpart AA.

The development and publication of these rules for the construction industry took many years, and in fact the struggle to come to a consensus on what is needed, versus what is feasible and what is economically viable, probably serves to illustrate just how much this Standard was needed. There was (and still is) much disagreement among different segments of the industry, facility owners, employers and labor representatives as to what should or even could be included in a set of rules for the construction industry. The previous lack of any sort of common practice in the construction industry probably underscores the importance and the need for a set of rules for the industry to follow.

I have spent my entire working life employed in one capacity or another in the construction industry. One thing I know to be true is that this industry is not one that quickly embraces change. This book is being written a year after the release of the new Construction Industry Standards for confined spaces. At this time, my opinion is that there are many companies in the construction industry that have not even heard of these new rules. There are probably many others that know they were published but assume that they don't need to concern themselves with them at this time for one reason or another, and there are probably still more that know about the new rules and just don't have any idea how they will be affected or what they need to do. I think that the number of construction companies that are already actively seeking to incorporate these rules into their everyday work practices probably represents a small minority of construction companies.

I commonly get asked by contractors, "do these rules actually apply to me" and the answer is an easy "yes, they do". OSHA published the proposed rule in the Federal Register back on November 28, 2007, and then proceeded with the very lengthy and very public process of enacting a new standard. This culminated with the publication of the final rule in the Federal Register on May 4, 2015. The Standard took effect on August 3, 2015. OSHA has published frequently asked questions, fact sheets, quick cards, a small entity compliance guide, and has issued compliance directives. At the time of this book's publication, the rule is in full force and effect and is being enforced in all states and regions under the jurisdiction of Federal OSHA.

Unfortunately, none of this necessarily means that industry is complying, or even trying to comply. For one thing, OSHA issued, and then extended, a temporary stay on enforcement of this Standard in the residential construction industry. In spite of the fact that this was a very targeted directive that was meant to allow the residential construction industry more time to evaluate how this will affect every day operations like entering attics and other crawl spaces, I continue to hear and see statements to the effect of "OSHA issued a new rule, but it is not being enforced yet". Another issue is the State-Plan States. About half of the country is covered by some form of a State run OSHA, and these States have some time granted to them before they are required to adopt the new rule, or a rule that is "at least as effective as" the new rule. And finally, regardless of whether a company falls into one of these temporary delays of enforcement or not, many construction companies simply may not know about the new rules and some just do not comply with new rules until they are forced to do so.

This book is written to both help and encourage construction companies to start managing confined spaces on their project in a manner that is compliant with the Construction Industry Standards. The goal is to get contractors to understand the importance and relevance to both their companies and their employees. The Standard for Confined Spaces in Construction is long (it occupies about 27 pages in this book as Appendix B) and comprehensive. In addition to just defining permit-required confined spaces and covering the rules that must be followed to enter, it covers everything from the process of continual site evaluation to the responsibilities assigned to project owners, general contractors, and subcontractors, as well as the required communication that must occur between these parties. This book is meant to break down the rules in a manner that makes them easy to understand, and more importantly this book is written to help companies understand both the importance and the benefits of actively managing confined spaces in construction.

In evaluating the need for this new Standard, the United States Secretary of Labor Thomas E. Perez stated "This new rule will significantly improve the safety of construction workers who enter confined spaces. In fact, we estimate that it will prevent about 780 serious injuries every year."

"This rule will save lives of construction workers," said Assistant Secretary of Labor for Occupational Safety and Health Dr. David Michaels. "Unlike most general industry worksites, construction sites are continually evolving, with the number and characteristics of confined spaces changing as work progresses. This rule emphasizes training, continuous worksite evaluation and communication requirements to further protect workers' safety and health."

As I began my research into these rules and went through my process for breaking them down to create training and compliance packages for various clients, I came to the conclusion that while the Standard is comprehensive, it is also flexible. It does require a level of evaluation and communication that I believe was not previously very common in the construction industry, and that will probably have many companies in an uproar over new things that we must do. But I have also found that forcing this evaluation and requiring formal dialog between the project owner, the general contractor and the trade contractors has led my clients not only to a safer worksite, but one that is more productive. In some cases, the forced evaluation has led to collaborative sessions that have concluded not in how we are going to stay safe inside these spaces, but instead have led to a change in the sequence of work that has eliminated the need to work in a hazardous confined space. The results have been not only an improvement in the safe working conditions, but an improvement in the productivity on the site and a better understanding of working conditions for the trade contractors.

Chapters 2, 3 and 4 started out as a series of posts on LinkedIn. I wrote these articles and posted them shortly after the new Construction Industry Standard was written to try to help contractors break down the rules and figure what they needed to do to comply. These articles were very successful and received much praise for being straight forward and very effective in unraveling and explaining the rules without being overly complicated. If you are not yet familiar with the Standard, read these chapters all the way through before moving on to the details. This will quickly get you up to speed on what the rules are and what you need to do to keep people safe.

Subsequent Chapter's explore additional details of confined space entry and inspection in more detail, and compare the OSHA regulations to other rules and standards. They also discuss the practical impacts that these new rules could have on various segments of our industry, including infrastructure inspection, repair and maintenance and residential construction.

FIGURE 1 – AN OFTEN-OVERLOOKED CONFINED SPACE

Chapter 8 explores why the industry needed this rule and includes several case studies. These case studies illustrate not only the hazards that can be present in confined spaces, but also the fact that these hazards may not be obvious and might not even be present every time you enter. They also show that despite this, when hazards do arise in confined spaces, the results are often deadly.

The book concludes with some practical advice for contractors and introduces the concept of designing for safety. Discussion is included on how this concept can be applied to the management of confined spaces in the hopes that contractors who may have input during the design phase of a structure will take what has been presented here into consideration and make suggestions that will result in safer construction and safer future maintenance activities.

Appendix A contains a thorough discussion on the proper use of the equipment that's used to test and monitor the air inside a confined space. The attributes of many confined spaces in construction include the potential for a hazardous atmosphere, and in general industry testing the air prior to entering or even prior to classifying the space has become a standard activity. As this becomes the norm in the construction industry, it is critical that these pieces of equipment be used properly. This appendix discusses the calibration, testing and use of these devices.

Appendix B to this book contains the full Standard for Confined Spaces in Construction. A separate terms and definition section is not included because all the pertinent terms are defined directly in the rule itself in §1926.1202. The definitions given are specific to this rule and its application to the construction industry, so this is the best reference for the terms used throughout this book. These terms that are defined in the Standard are capitalized when used in the body of this text.

2 UNDERSTANDING THE RULES AND DEFINING THE HAZARDS

The rules published by OSHA are extensive and are a good example of using mandatory process requirements to keep people safe. As discussed in the Introduction, the lack of a mandatory set of rules in the construction industry has contributed to confusion, abuse, and even injuries and fatalities. In the U.S., during the five-year period from 2005 through 2009, there were a reported 481 fatalities that resulted from entry into hazardous confined spaces. This averages out to about 1 fatality every work day in the U.S. related to confined space entry. In about half of these fatalities the victim worked in the construction industry regularly and over 60% of the incidents occurred during construction related activities. OSHA used these statistics in claiming that there was a need for a new set of rules, and Chapter 8 cites several examples of fatal conditions that have occurred in the past and are likely to continue to be fatal unless the industry learns from its mistakes and takes proactive steps to protect its employees.

In choosing to draw up a new set of rules rather than just adopting the existing General Industry Standard, OSHA recognized the fact that construction sites are frequently very different places than fixed general industry sites. For one thing, the new rules recognize the varying roles played by multiple parties on many construction sites, and the new standard does a good job of assigning roles and responsibilities on these sites with a general contractor and multiple sub-contractors. The standard recognizes the fact that many construction workers do not understand some of the hazards that can exist and can be unique to confined spaces. It also recognizes the fact that many employers and their employees may not even know of the existence of these hazardous spaces at some of their work sites, and it requires proactive steps on the part of all construction employers to correct this lack of knowledge.

The new standard uses the same definition of a confined space as the general industry standard; i.e. large enough to enter, not designed for continuous occupancy and has limited means for entry and exit; however, this is the first time the construction industry has been presented with a precise definition of a permit-required confined space and given a mandatory set of compliance rules. The new standard also goes much further than the old language, contained in 29 CFR 1926 Subpart C (general safety and health provisions), in giving examples of what is to be considered a confined space... elevator pits, manholes under construction, concrete pier columns, vaults, enclosed beams, crawl spaces, attics and HVAC ducts are all mentioned in the scope of this new standard.

OSHA has issued several documents that give examples of spaces that potentially meet the definition of a confined space. These include...

■ Manholes (such as sewer, storm drain, electrical, communication, or other utility) ■ Sewers ■ Storm drains ■ Water mains ■ Lift stations ■ Tanks (such as fuel, chemical, water or other liquid, solid or gas) ■ Pits (such as elevator, escalator pump, valve or other equipment) ■ Bins ■ Boilers ■ Incinerators ■ Scrubbers ■ Concrete pier columns ■ Transformer vaults ■ Heating, ventilation, and air conditioning (HVAC) ducts ■ Precast concrete and other pre-formed manhole units ■ Drilled shafts ■ Enclosed beams ■ Vessels ■ Digesters ■ Cesspools ■ Silos ■ Air receivers ■ Sludge gates ■ Air preheaters ■ Transformers ■ Turbines ■ Chillers ■ Bag houses ■ Mixers/reactors ■ Crawl spaces ■ Attics ■ Basements (before steps are installed).

To begin with, the standard now specifically requires that employers ensure that a competent person identifies all confined spaces in which one or more of the employees it directs may work, and must identify each of those spaces that is a permit-required space. This must be done prior to beginning work at a job site. Additional roles and responsibilities are specifically defined and assigned throughout the lengthy standard. Roles defined under this rule include the following:

- **Host Employer** - The employer that owns or manages the property where the construction work is taking place - this means the owner of a site has responsibilities under this new rule, even if they hire a controlling contractor

- **Controlling Contractor** - The employer that has overall responsibility for construction at the worksite, for example the general contractor (if the controlling contractor owns or manages the property, then it is both a controlling employer and a host employer – for example a home builder)

- **Entry Employer** - Any employer who decides that an employee it directs will enter a permit space

It is interesting that this standard goes as far as to specifically define the owner of a construction site as a Host Employer. Since this Standard assigns specific responsibilities to the Host Employer it makes a project owner responsible for compliance with parts of this rule. A property owner or property manager is specifically permitted to contract away these obligations to a general contractor (or other Controlling Contractor) if it communicates this intent and transfers information regarding known confined spaces over to the contractor. This relieves the owner of responsibilities while the contractor is under control of the job site. It also makes the general contractor both the Host Employer and the Controlling Contractor. OSHA acknowledges that there will only be one Host Employer for the purposes of this standard, so it is advisable for the owner to include implicit language in their contract documents with their Controlling Contractor. Remember, if the owner remains the Host Employer, they can be cited along with the Controlling Contractor for any violations; and regardless of the assigned role, the owner must communicate information on existing confined spaces.

The definitions section of the Standard also contains the following note:

> *An employer cannot avoid the duties of the standard merely by refusing to decide whether its employees will enter a permit space, and OSHA will consider the failure to so decide to be an implicit decision to allow employees to enter those spaces if they are working in the proximity of the space.*

Combining all the items discussed so far, this new standard can be seen as having a significant effect on how potentially every construction site is managed in the future. Unless proactive steps are taken on virtually all construction sites, by all employers, the owner of the project potentially becomes a Host Employer, and all contractors and sub-contractors on the site could potentially be seen as being in violation of the standard if their employees are merely working in the proximity of a permit required confined space without having discussed confined space hazards with their employees.

While this may seem harsh and over reaching at first glance, OSHA has created this standard not only in response to fatalities of workers in confined spaces, but in response to the circumstances surrounding many of these incidents. Because of the nature of many construction sites, there now exists an implicit requirement for contractors to train practically all employees on how to recognize a confined space and what the hazards of entering those spaces can be. Even contractors whose employees are not expected to work in permit required confined spaces need to now make sure they teach those employees about the hazards of these space, and they need to incorporate a policy that specifically prohibits those employees from entering them. Again, at first glance that may seem a bit over reaching; however, considering the percentage of fatalities that occur when onlookers rush into confined spaces to try to rescue someone, this simple bit of instruction may save someone's life.

Most of the requirements for additional training, personnel, equipment and specialized procedures required under this rule only occur when a confined space that requires entry is determined to be a permit-required confined space. A space generally becomes permit-required when it poses additional hazards to the entrants that could subject them to injury and/or make it difficult for them to exit or be rescued. The standard is aimed at identifying these types of spaces and requiring additional steps to control the hazards and provide for timely and effective rescue in the event of an incident. Specifically, the Standard defines a permit-required confined space as a confined space that has one or more of the following characteristics:

1. Contains or has a potential to contain a hazardous atmosphere

2. Contains a material that has the potential for engulfing an entrant

3. Has an internal configuration such that an entrant could be trapped or asphyxiated by inwardly converging walls or by a floor which slopes downward and tapers to a smaller cross-section

4. Contains any other recognized serious safety or health hazard.

Once a space is characterized as permit-required, it mandates significant additional assessments to determine appropriate measures that will need to be taken to protect Entry Employees. These procedures must be documented in a written Permit procedure that will dictate entry requirements, including the need for Attendants, an Entry Supervisor and Rescue Personnel.

An example of compliance with this new Standard can be summarized as follows:

- A project owner begins by making a decision to remain the Host Employer, or to transfer that responsibility to a Controlling Contractor.

- Remember that there can only be one Host Employer, so if an owner is hiring multiple trades and managing them without a general contractor, it is doubtful that the owner can contract away their role as a Host Employer under this Standard.

- Either way, the owner must identify all known confined spaces, note them as permit-required or not, and communicate any precautions that they or any previous

Controlling Contractor or Entry Employer implemented for the protection of employees in any Permit spaces.

   o Note that at a site that is operating under the General Industry rules, that standard also requires the owner to identify and label confined spaces

- The owner should clearly communicate this transfer of responsibility in the contract documents.

- In new construction, with no existing facilities, this is probably a simple process whereby the owner notes (in the Contract Documents) that there are no known confined spaces on the property and that responsibility as the Host Employer is transferred to the general contractor for the duration of the project.

- A general contractor or construction management firm that will become the Controlling Contractor should implement a written policy for confined spaces, and should mandate compliance with that policy for all its employees and subcontractors. Subcontractors should be required to analyze their work and provide notice of any confined spaces that it may create or enter during the course of construction.

- At a minimum, the Controlling Contractor should train all its employees on the recognition and hazards of confined spaces, and needs to decide if its employees will be permitted to enter permit-required confined spaces. If they will be permitted to enter these spaces, then additional training will be required.

   o This includes supervisory personnel and inspection personnel – see additional discussion in Chapter 4

- The Controlling Contractor must communicate knowledge of any known confined spaces on the project to all subcontractors.

- If the Controlling Contractor identifies, or receives notice of, a permit required confined space they must do the following:

   o Post appropriate danger signs

   o Communicate the danger to all employees in an additional manner other than posting

- On some projects, this may be an ongoing process. As new trades begin work and provide notification, this will need to be communicated with all other personnel on the project.

- The Controlling Contractor is required to coordinate multiple entries or entries by multiple subcontractors and is required to debrief each Entry Contractor after entry is completed to assess hazards actually encountered and effectiveness of the procedures.

- All trade contractors will need to train their employees on the recognition of confined spaces and the hazards they create. They will need to establish a written policy for entry into permit-required confined spaces.

- Even if a trade contractor's work does not typically require employees to enter permit-required confined spaces they should provide and document this awareness training, and establish a written policy that prohibits entry unless specifically authorized.

- All trade contractors should establish a proactive system of notification to their employees. As the contractor receives notifications of permit-spaces on job sites where their employees work, they need to take proactive steps to communicate this information to their employees at the site(s) and document this communication. Again, if entry will not be required this can be as simple as a pre-shift orientation meeting where the crew leader informs all employees that they have been notified of a new permit-space on the job site and no one is authorized to enter this space for any reason. Make sure to document this meeting!

- If a trade contractor's work will require them to enter any of these permit-spaces, they become an Entry Contractor and further action is needed as follows:
    - They must have a written permit-space program that complies with the requirements of 29 CFR 2926.1204. This written program must be available on-site prior to, and during, entry operations.
    - They will be required to provide and document additional training in conformance with 29 CFR 1926.1207 related to permit-required confined space entry.
    - The Entry Contractor's Competent Person must assess and address the site-specific hazards and provide this documentation to the Controlling Contractor.
    - Any employee that will become an Entrant must be provided with appropriate testing, monitoring, lighting, ventilation, communications, and other personal protective equipment that is needed.
    - The Entry Contractor must secure the space from entry by others and ensure compliance with permit procedures.

o Entry will require that the Entry Contractor provide entry supervisors, attendants and rescue personnel as determined by the hazard assessment and as discussed in subsequent chapters of this book.

o The Entry Contractor becomes responsible for controlling all hazards to its employees who will work on the entry team.

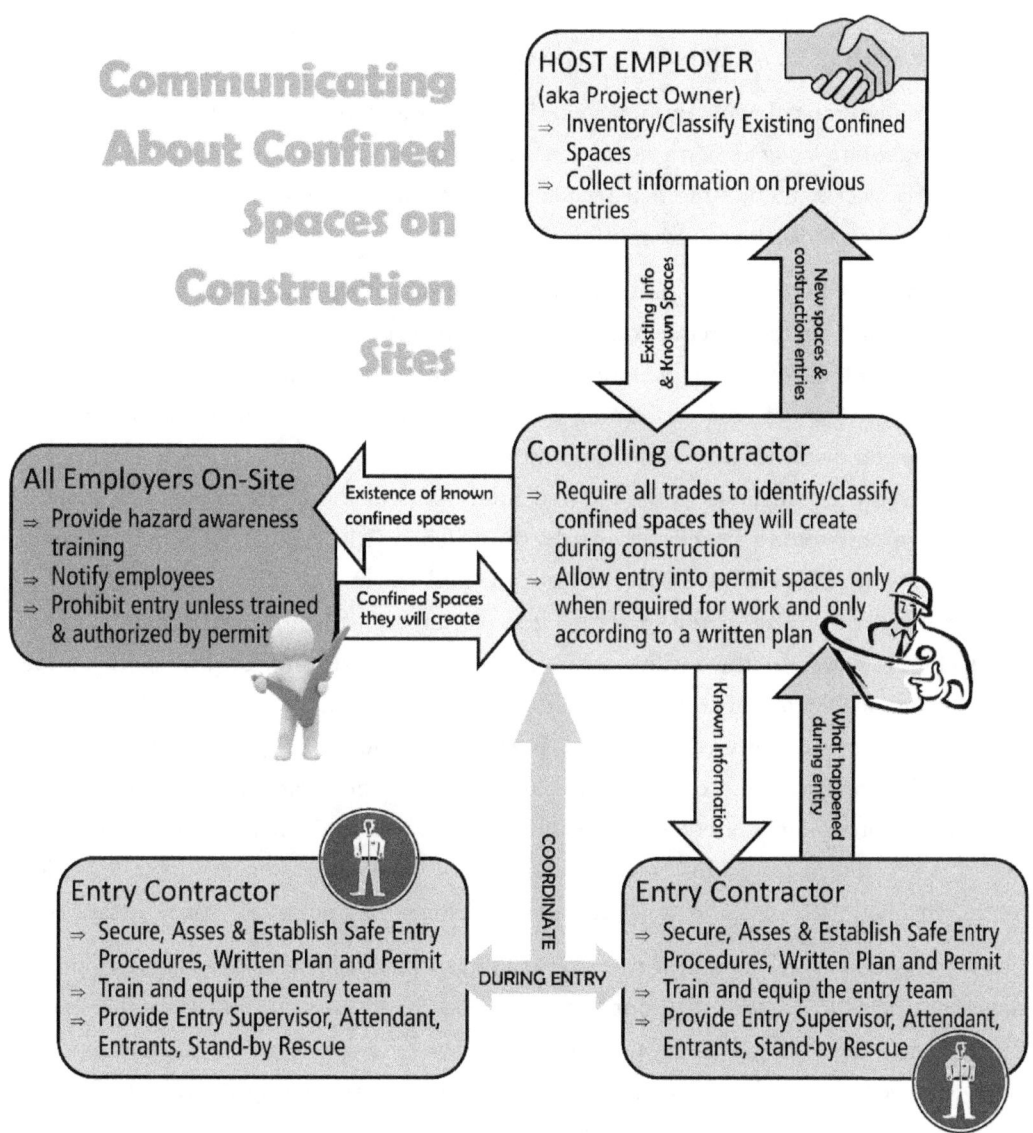

FIGURE 2 – COMMUNICATIONS FLOWCHART

There are several options available to an Entry Contractor who must perform work in a permit-required confined space, and these are discussed in detail in subsequent chapters. Contractors are advised, however, that there are times when additional discussion with all parties involved in the project may provide insight as to how the work can be performed without entering a hazardous confined space. For example, resequencing the work on the project could lead to a schedule that has work being conducting after all hazards are mitigated, or even after a space is opened up and no longer fits the definition of a confined space.

AUTHORS OBSERVATIONS – A LESSON LEARNED IN COMMUNICATION

The communications flowchart shown in Figure 2 does little good if the parties do not realize their roles. While at first glance, these assignments may seem obvious, a recent near miss on a project reminded me that sometimes construction projects can get complicated, and without formal discussion, assumptions can be made that lead to errors…

My client, a general contractor, was performing construction work at an existing plant facility where the owner exerts extreme control over the management of the project due to the critical nature of the ongoing operations and activities at the facility. There was also an additional general contractor performing unrelated work at the site at the same time, and towards the end of my client's work, the plant was going into shut down mode and bringing in numerous other companies to the site to conduct work, some of which could be argued is maintenance and some of which is clearly construction. My client's subcontractors were performing work in a confined space using a method where they eliminate and proved no hazards in order to reclassify the space to a non-permit space. A near miss incident occurred in which the power in the facility failed, leaving the entrants inside the space in total darkness and unable to move a powered overhead structure being worked on, which resulted in blocking their primary means of exit. My client and their team performed very well in immediately terminating the entry and invoking an emergency plan that we had previously worked up. Interestingly enough, at the time I walked through this emergency plan with them I believe they all thought it inconceivable that it would ever be needed. None the less, the entrants had flashlights and an alternate means of climbing out of the space unharmed.

The investigation revealed that the power outage was a planned and scheduled event, but my client and their project team had not been notified. The event was further complicated by the fact that the power outage and planned cut-over did not go as planned and the emergency generator did not engage for a variety of reasons. During this event, my client and their team were the only personnel that remained in the facility as the owner had informed all other contractors and their own employees of the scheduled power outage and relocated them to a space outside the facility while the power cut-over took place.

Among my conclusions were the fact that the Controlling Entity had failed to effectively coordinate the work of the multiple contractors on site, and failed to notify my client that they would be altering the conditions on which our subcontractors were relying in making a determination that no hazards existed. That simple statement drew immediate disagreement and some confusion from all parties involved. Upon discussion, I realized that the various parties were not recognizing the facility owner as the controlling entity, and therefore my client (a general contractor) was assuming that I was stating they failed to properly coordinate something they knew nothing about. The owner's representative initially disagreed with my classification of them as the controlling entity and was therefore reluctant to accept responsibility for failing to coordinate activities. And of course, the entry contractor was also dismayed and just wanted to know why someone shut off the electricity while they were in the space.

I quickly realized that there was a failure to properly communicate, and that failure could be traced all the way back to the beginning of the project and the discussion and assignment of roles and responsibilities. This was a big reminder that in construction every project is different, and making generalizations and assumptions can get you in trouble. Some projects are straightforward...on a new construction project or even a renovation or tenant buildout where the entire site is turned over to the general contractor, it is easy to define roles. There is one general contractor and multiple subcontractors. The owner as the Host Employer fulfills its obligations by identifying known spaces and turning control over to the general contractor. The general contractor takes the role of the Controlling Entity and coordinates the activities of its subs. But there will also be cases like this example, where there are multiple general contractors, with different tiers of subcontractors, each performing work under the control of a facility owner while working alongside that facility's employees and maintenance personnel. This is where communication becomes crucial and it is important not to assume that each party knows the role they are supposed to play.

It is also important to understand that the rules for confined spaces in construction come with their own set of definitions and assignment of responsibilities. In my example, the owner is clearly the controlling entity with respect to confined space entry. They needed to recognize their role and needed to coordinate the plant's ongoing activities and multiple construction projects with my client to ensure that work that is clearly beyond the control of my client does not affect the safety of the confined space entrants. That does not necessarily mean that I consider them to be the controlling contractor with respect to the general Rules of Construction (that's a different discussion), but they clearly meet the definition of the Controlling Entity in the confined spaces Standard and that means they needed to take on the role and assume the responsibilities assigned to that entity. However, if that entity doesn't even recognize their role, it is unlikely that it will perform the assigned duties. The lesson learned here is that anytime entry into confined spaces is planned, discussion needs to take place in advance to ensure that each parties' roles are identified so that expectations are met and managed. In this case the facility manager had no idea that he would potentially impact the safety of the entrants, or that my client was relying on him to coordinate these efforts.

Communication must take place to ensure that roles and responsibilities are understood, and the parties that are the experts need to take the time to communicate their needs and expectations. When one of the parties to the operation is not knowledgeable about construction, it is critical to take the time to explain the procedures and the responsibilities so that they understand how seemingly unrelated work by others could affect the safety of your entry team. When there are multiple projects on-site, particularly those overseen by multiple companies, each of whom are typically seen as a controlling entity (such as a general contractor), or when there is a lack of an entity that would typically be identified as the controlling entity (as in the case of a facility owner directly hiring multiple subcontractors), it is critical to have this discussion up front to make sure assignment of roles are understood by all parties. Since this regulation assigns duties to each role, the safety of the entrants depends on those roles being recognized and duties performed.

In digging further into this near miss incident, I determined that the root cause was a failure to communicate (not shocking on a construction project), but it was a failure to communicate by virtually all parties. The entry employer certainly realized that safe entry was dependent upon power being supplied, but this important detail was not specifically communicated to anyone. The general contractor on the job (as well as myself as their consultant) did not specifically communicate their expectations to the plant management team related to the fact that we were considering them to be the Controlling Entity with respect to confined space entry operations. Finally, the facility owner did not recognize the fact that since they were maintaining control of the operations at the plant, including work being conducted by multiple general contractors and maintenance personnel, they were the Controlling Entity with respect to the confined space entries taking place on-site. To put all of this in simpler terms; the entry contractor assumed that it was obvious that electricity was needed to maintain safe entry conditions, and we all assumed that the plant would notify us if they were going to shut off power to the entire facility while we were inside.

The lesson learned here is that basic communication is critical to the health and welfare of people working in confined spaces. These spaces are unique, especially on construction sites where conditions can change in an instant, and these changes can be set off by parties conducting work that is seemingly completely unrelated to the work being conducted in the confined space. Because of this, we owe it to the confined space entrants to take the time to set aside a special discussion with all parties when we determine that work on a project requires entry into a confined space. Make sure this meeting includes a discussion on the assignment of roles and responsibilities, a discussion about any assumptions being made or conditions being relied upon to control or eliminate hazards, and an outline of the entry procedures and emergency protocols that have been developed. In addition, it is always advisable when conducting these types of meetings to insist that the proper people be present. In other words, the actual people that will be on-site, not just any company representative that may or may not pass-on the critical information.

3 CLASSIFYING PERMIT REQUIRED CONFINED SPACES

The definition of a Permit Required Confined Space in construction is similar to the General Industry Standard. The new 29 CFR 1926.1202 standard states that a permit-space is any confined space having one or more of the following characteristics:

1. Contains or has a potential to contain a hazardous atmosphere

2. Contains a material that has the potential for engulfing an entrant

3. Has an internal configuration such that an entrant could be trapped or asphyxiated by inwardly converging walls or by a floor which slopes downward and tapers to a smaller cross-section

4. Contains any other recognized serious safety or health hazard

Where the construction standard begins to differ from the general industry standard is in how a space gets classified, and by whom. Since construction sites by nature involve an ever changing facility occupied by multiple employers (subcontractors), the standard assigns responsibilities for identifying and classifying confined spaces to several entities. This is explained in detail in Chapter 2, and summarized as follows:

1. The Host Employer and/or Controlling Contractor must identify all known existing confined spaces prior to work commencing on a job site

2. They must designate each of these spaces as permit-required or non-permit spaces

3. Danger signs must be posted at each of these permit-required spaces and sub-contractors must be notified of their existence

4. All contractors / sub-contractors on site must determine if their work requires entry into one of these spaces

a. If entry is not required, each employer must still take steps to teach their employees about the hazards and the reasons for not entering, and must prohibit its employees from entering these spaces

b. If a contractor's work does require it to enter one of these spaces, they become an Entry Contractor and they must have a competent person investigate the space and determine the hazards that will exist so they can develop a means of proper and safe entry

5. As a construction project progresses, new confined spaces may be constructed, requiring continuous identification and marking along with updated notifications

FIGURE 3 – CONFINED SPACE SIGNAGE

Once a confined space is designated as a permit-required space by the Host Employer, the Controlling Contractor or an Entry Contractor, entry must be controlled by a written Permit program operated by the Entry Employer. Note that there are no exceptions to this rule. Once the "permit-required" label has been applied to a confined space, the rules of this new standard must be followed; i.e. there must be a written plan and Permit document, and entry can only be made under one of the three scenarios described herein. Even reclassification, or downgrading a space to non-permit because hazards have been eliminated, requires actions to be documented and certified.

ENTRY METHOD 1 – RECLASSIFICATION

Since construction worksites are inherently evolving and conditions can change on a regular basis, the construction standard allows for the reclassification or downgrading of a confined space if the following conditions are met:

- All hazards that were identified when the space was classified as permit-required must be either eliminated or isolated through engineering controls – this may include de-energizing and locking out circuits or equipment or blanking, blocking or misaligning pipes to eliminate engulfment hazards during entry

- The work that will be performed in the space cannot create any additional hazards, such as hot work or applying coatings that could result in a hazardous atmosphere

- There can be no actual atmospheric hazard present in the space

- There can be no potential for an atmospheric hazard to develop – note that if ventilation is required to control the atmosphere and make it safe, the space cannot be downgraded to a non-permit space; instead, consider the alternate entry procedure #2 discussed herein

The requirement to control hazardous energy also introduces another program, known as Lock-Out / Tag-Out (LOTO), which has not previously been included in the Construction Industry Regulations. While the General Duty Clause has always mandated that contractors do something to control hazardous energy, the construction industry has not typically faced the more stringent and formal requirements of the General Industry's LOTO program required under 29 CFR 1910.147 (General Industry Regulations). While this is probably still subject to future interpretation and enforcement guidelines, contractors are best advised to establish a written LOTO program that complies with the requirements contained in the General Industry Standards any time hazardous energy needs to be controlled in a confined space. In general, this includes developing a system where each exposed employee (in this case each Entrant) has their own lock and tag applied at the point of control to ensure they are each protected.

To reclassify a space, a Competent Person representing the Entry Contractor must verify all of the above actions have been taken to render the space safe and justify its reclassification. While the standard is not specific on how to handle and maintain the paperwork under these conditions of reclassification, it does specifically state that the Entry Contractor must provide written documentation of what was done to eliminate or control the previously identified hazards. It also states that the documentation must be dated and signed by the Competent Person who is certifying that the space is safe, and that the documentation must be made available to all Entrants.

DON'T FORGET LOTO!

Lock Out Tag Out, or LOTO, is a procedure whereby formal steps are taken to ensure that systems that release energy are temporarily rendered harmless while working in a confined space. Energy may be present in a space in many different forms and all of them should be considered when evaluating the hazards of a confined space. The form of energy that comes to most people's minds when discussing a LOTO program is electrical energy, and the most common way to control that energy is to shut it off at a breaker or disconnect switch. When evaluating hazards in a confined space, it is also important to consider other sources of energy such as mechanical energy presented by the unexpected start-up of a piece of machinery with unguarded moving parts. A proper LOTO program communicates the fact that it is not acceptable to simply turn off these energy sources. They must also be "locked out" to ensure they are not accidentally restarted during work operations. Those that already utilize a formal LOTO program need to be aware that while other regulations allow the use of a tag to warn others not re-energize a system, the rules for confined spaces dictate that the entrants use an actual lock unless it is infeasible (§1926.1202). It is also important to remember that some systems store energy even after they are disconnected and the lockout procedures must include a means for dissipating that energy and testing the system to ensure it is de-energized and safe. System specific LOTO procedures should be listed on the Permit form and stand-by rescue personnel need to be aware of any systems that have been temporarily de-energized and locked out so that they can ensure their own safety in the case of an entry rescue.

With the ever changing nature of construction sites and the typical multi-employer nature of most of these sites, it is advisable to approach any reclassification as temporary and specific to a single entry employer. In other words, develop a Permit form that contains fields for documenting the reclassification. Approach the space first as permit-required until the hazards are eliminated or isolated. Document these actions on the Permit, and then mark the Permit form as "Temporarily Reclassified to Non-Permit Entry". Include the date of entry and the certification by the Competent Person, and cancel the reclassification when entry operations are complete. Approaching the reclassification in this manner leaves a documented trail of the formal steps that were taken to render the space safe and justify its reclassification. Making the reclassification temporary and specific to each entry operation ensures that the space remains identified as hazardous and warns others not to enter the space. It also ensures that the permit-required designation stays in place when your entry operations are completed and you remove any temporary isolation or controls that were put in place.

Remember that all work performed to render the space safe must be conducted from outside the space, including any testing or inspection required to evaluate the space. If entry is required to perform this work, the permit-required entry procedures must be followed until the space is officially reclassified. Also, keep in mind that using mechanical ventilation to maintain a safe atmosphere does not constitute an elimination of the hazard under this section and you cannot reclassify the space under this condition. Instead, evaluate the alternate entry procedure discussed next.

ENTRY METHOD 2 – ELIMINATION OF ALL HAZARDS EXCEPT ATMOSPHERIC

The construction standard allows a modified type of permit-required confined space entry that can be utilized when all hazards are eliminated, except for the possibility of hazardous atmospheric conditions. Remember, any confined space that "contains or has the potential to contain a hazardous atmosphere" must be designated as a permit-required space. The inclusion of the phrase "has the potential to contain" is significant in that it pulls in many different types of spaces that are subject to potential infiltration by gasses such as carbon monoxide (from generators, construction equipment or motor vehicles) and hydrogen sulfide (from adjacent sewer systems), and spaces that might be oxygen deficient (such as utility vaults). While the mere potential for a hazardous atmosphere mandates the classification of a space to permit-required, this option can be utilized to minimize the impact to cost and productivity by following a written plan designed to increase safety and minimize risk.

The use of this option requires the Entry Contractor to meet these requirements:

- All hazards except for the potential hazardous atmospheric conditions must be eliminated or isolated in the same manner as prescribed under the re-classification method

- The atmosphere inside the space must be properly tested to confirm safe conditions prior to making entry

- Continuous mechanical ventilation must be maintained during the entry operations

- The atmospheric conditions must be continuously monitored during entry operations

- Entry must be terminated if mechanical ventilation cannot maintain safe atmospheric conditions or if any other unanticipated hazards are encountered during entry

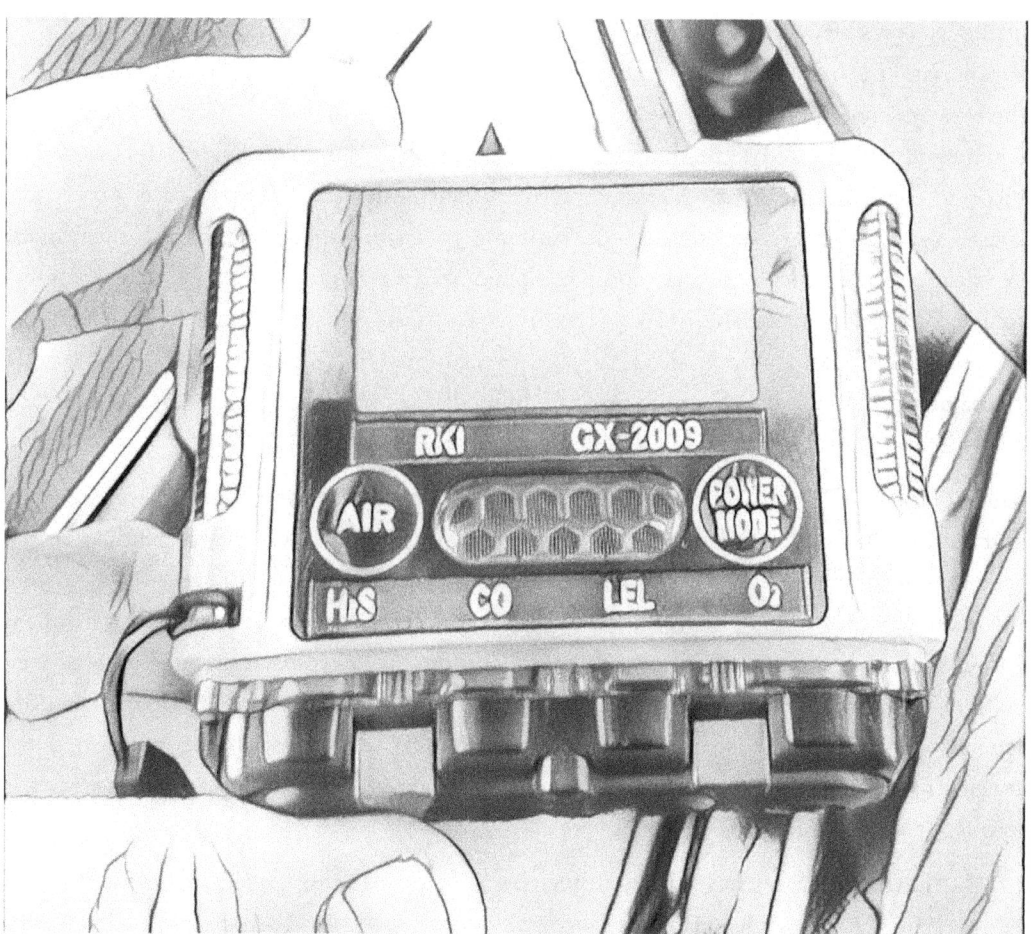

FIGURE 4 – FOUR GAS ATMOSPHERIC MONITOR

These actions, steps taken to eliminate hazards, and atmospheric test results must all be documented by the Competent Person representing the Entry Contractor and all of this information must be made available to everyone on the entry team; therefore it is best to approach this procedure in the same manner as described under the reclassification procedure, i.e. use the Permit form to document the steps taken for a specific entry, note the entry as being done using the modified procedures of 29 CFR 1926.1293(e), and cancel the written permit at the conclusion of each entry operation. This modified method of entry into a permit-required space results in safer working conditions for the entrants and pays off for the contractor by eliminating the requirements for maintaining a dedicated attendant, and on-site stand-by rescue equipment and personnel.

The use of this modified entry procedure mandates that the Entry Contractor utilize two specific types of equipment: an atmospheric monitor to test and continuously monitor air quality, and a mechanical ventilation fan with ducting as needed to ventilate the space. Additional information on monitors and equipment used to test the atmospheric quality inside of a confined space is included in Appendix A. Mechanical ventilation fans should be appropriate for their intended use and need to be sized to provide adequate ventilation given the size and configuration of the space. It is important to note that like any mechanical equipment, there is always the possibility of breakdown or equipment failure. If a decision is made to eliminate the Entry Attendant as permitted under this alternate entry procedure, special care needs to be taken by the Entrants to ensure they monitor the mechanical ventilation. Some equipment may come equipped with an alarm that can alert the Entrants to its failure, but even so, absent an Attendant to watch the equipment and alert the Entrants to any failure, the Entrants need to take precautions to recognize any failure of the ventilation equipment which would mandate immediate evacuation until repairs can be made or equipment can be replaced.

The Competent Person will be required to use an appropriate gas meter to test the space without making entry. The exact method for conducting this test will vary depending on the space and its configuration, but affordable monitors can be purchased or rented to complete this requirement. A common type used to detect conditions anticipated on many job sites is a four-gas meter that tests for oxygen, carbon monoxide, explosive gases and hydrogen sulfide. This covers the risks most commonly associated with the types of spaces that are entered on construction sites; however, additional tests may need to be performed for specific gasses if there is another known hazard. When checking atmospheric quality, it is important to remember several things:

1. Different gasses have different weights – you must check at least the top, bottom and middle of the space to ensure that you detect gasses that are lighter, heavier or the same weight as the air

2. Active sewer systems are not the only places that can contain hydrogen sulfide – this gas, which is commonly generated in sewer lines, can travel through the soil when it escapes through cracks or leaks in the system

3. Carbon monoxide is a colorless and odorless gas that is created by the exhaust of generators, equipment and cars – simply parking a running truck adjacent to the hatch of a utility vault can flood the space with deadly carbon monoxide

4. Oxygen in a space can be used up by things like organic growth (mold), open flames, or even corrosion of metal parts

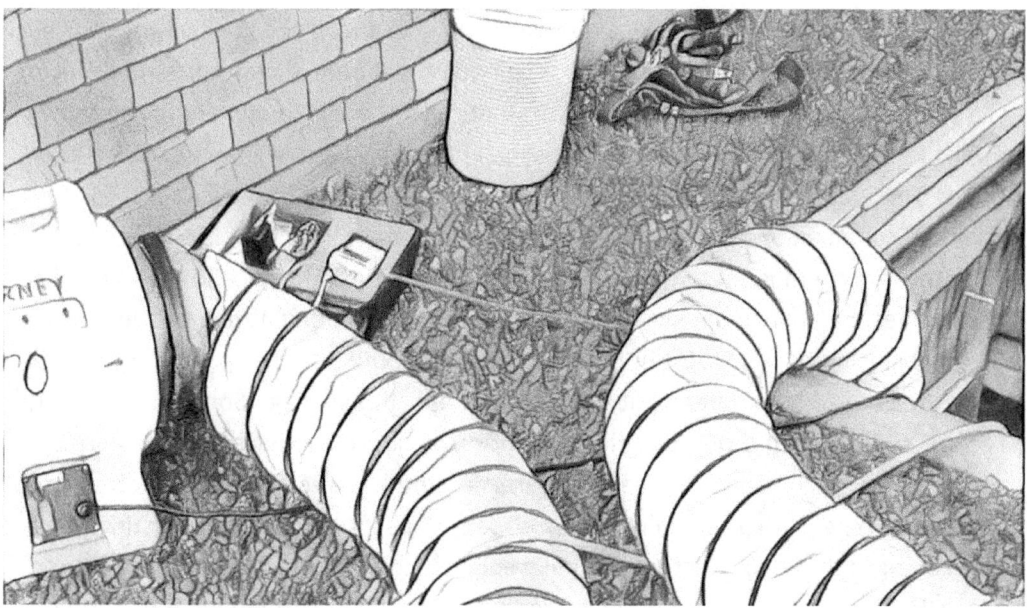

FIGURE 5 – MONITORING AND VENTILATION

Remember that testing must be performed without making entry. Most equipment that is commonly used on construction sites today is very small and can be tethered and lowered into a space, attached to a pole to push it horizontally into a space, or it may come with a sampling pump and length of tubing that allows the meter to stay with you while you lower a sampling probe into the space. This is especially useful if the space may contain water that could damage the testing equipment.

Persons conducting this testing should be trained individuals who understand the use and limitations of the testing equipment, and who have knowledge about the space to be entered, the potential hazards, and the work to be performed inside. After testing the space and verifying safe entry conditions, the results of the tests are recorded on the Permit form along with any other steps previously taken to render the space safe. When the entrants make entry into the space, they must bring the equipment with them to continuously monitor conditions in the space. When calibrated, set-up and used properly, the atmospheric monitor will alarm to warn the entrants of changing conditions. Such an alarm would be an indication of deteriorating conditions and is a signal for immediate evacuation. Do not attempt to determine the cause while inside the space. Exit to safety and then asses the conditions from outside the space to determine what is happening.

CONTINOUS ATMOSPHERIC MONITORING VERSUS PERIODIC TESTING

As with many things, technological advances have changed the way we monitor air quality. Not only has the cost of this testing equipment gone down substantially over the years, its portability has improved dramatically as well. Equipment is commonly available that is not much larger than the cell phone you are probably carrying at all times. They are equipped with rechargeable batteries that can easily last all day. They can sound an audible alarm, flash lights and even vibrate (just like your phone). With these technological advances, there can be very little justification for selecting periodic testing instead of continuous atmospheric monitoring. When the device can be simply clipped to your vest or belt and taken with you into the space, it is going to be difficult to prove infeasibility for continuous monitoring. Some spaces may require additional consideration though. For example, some spaces may require the use of "intrinsically safe" devices that are designed to reduce the risk of sparks, explosions or electrical conductivity, and some spaces may require testing for a particular gas that necessitates additional specialized equipment. For these reasons, the OSHA Standard does allow for periodic testing when continuous monitoring is infeasible, but keep in mind that justification of infeasibility cannot include stating that you don't have enough meters to take into each space. These devices can typically be rented for a very reasonable price and training a person on its proper use is not difficult. These are simple steps to take to keep someone safe and alive. See Appendix B for more information on these devices.

It is not uncommon for the initial test values to indicate an unsafe atmosphere that requires mechanical ventilation to bring the space to a habitable condition. Ventilation equipment is readily available for sale or rent and consists of a forced air fan and length of air duct. Proper ventilation is typically performed by lowering the duct to the bottom of the space (or pushing it to the back of the space) and keeping the actual fan outside the space. The concept is to push fresh air from outside the space, into the bottom or far reaches of the space to displace the bad air and push it out. Mechanical ventilation for 30 minutes or more may be required to flood the space with breathable air, and the ventilation will need to continue for the duration of the entry to ensure adequate fresh air. To ensure proper test results and safe working conditions, stop the mechanical ventilation before taking new readings in the space. Verify through testing that the space can be brought to safe conditions prior to entry. You should be able to ventilate the space, turn off the ventilation, take new readings and verify safe and stable conditions prior to turning the ventilation back on and making entry. This ensures that Entrants will not be immediately overcome if the ventilation fails.

Hazardous atmospheres cause the deaths of both untrained entrants and untrained rescuers every year – the standard contains the phrase "has the potential to contain" because it is so critical that these spaces be tested. Numerous fatalities each year can be attributed to entering an untested space and then being asphyxiated by a poisonous gas or lack of oxygen, and many more fatalities can be attributed to untrained would-be rescuers who rush in to help the victim who is unconscious. That is why advanced testing is mandatory prior to entry, and continuous monitoring is necessary while people are inside the space. If the gas monitor alarms, it is giving notice that conditions are deteriorating and immediate evacuation is required. Likewise, if the atmosphere cannot be rendered safe and stable, entry under this modified procedure is not proper – use the full permit-required procedure and evaluate the need for self-contained breathing gear (SCBA).

ENTRY METHOD 3 – STANDARD PERMIT ENTRY

If the space contains serious safety or health hazards that cannot be eliminated or isolated so that it can be reclassified according to one of the two methods previously described, all of the rules for permit-required confined space entry must be followed. This is a time consuming and costly entry procedure that is meant to reflect the level of the hazards involved and is mandated to ensure the safety of the entrants. Just like the use of personal protective equipment (PPE) should be considered the last line of defense, the use of the permit entry procedure should be utilized only after consideration has been given to how hazards could be isolated or eliminated. Doing so reduces danger to entrants and may justify the reclassification of the space or the use of modified entry procedures previously described.

There will be times when this type of full permit-entry must be performed. A space that has a configuration that could cause an entrant to become trapped or a space that has the potential for an entrant to become engulfed are examples that will require a permit-entry unless those physical attributes are altered. Other examples would include a space containing a hazardous atmosphere that cannot be controlled and requires the use of self-contained breathing gear, or spaces that contain hazardous moving or energized equipment that cannot be shut off and locked out.

In addition, there are times when it is the actual work to be performed in the space that introduces the hazard. Entry into a space to perform welding, or work that requires the use of a cutting torch, would mandate the permit-required classification and all the standard permit-entry procedures would have to be followed; as would entrance into a utility vault by a qualified person to conduct work on energized circuits or equipment. Other hazards to consider when classifying the space and planning the entry would include extreme heat or cold. Entry into a metal silo for an extended period with outside temperatures of 110 degrees could easily represent an additional hazard that would cause the space to be permit-required, while scheduling the same work at a time when the temperature is 60 degrees may eliminate that hazard.

A permit-entry requires trained entrants, a dedicated attendant, an on-site entry supervisor and a means for immediate rescue should something go wrong. This complete procedure and its requirements are discussed in the next Chapter.

SUMMARY OF ENTRY METHODS

At most worksites, the Construction Industry has not been faced with a mandatory set of rules for classification of permit-required confined spaces, nor have we been faced with a mandatory set of rules for entry procedures. In many situations, work being performed by contractors on existing facilities has been classified as construction, and the General Industry rules for confined spaces have not been followed. Previous Construction Industry rules have only contained brief language requiring workers to be trained about the hazards of confined spaces. This has now all been changed with the publication of the new 29 CFR 1926 Subpart AA – Confined Spaces in Construction. There are no longer any loopholes for employers whose employees enter confined spaces, regardless of whether you call it maintenance or construction. Any space that fits the definition of a confined space must be designated as permit or non-permit, and the appropriate procedures must be followed. For construction work performed on existing facilities, OSHA has stated that a contractor will be deemed to be in-compliance if they are following the new Construction Industry Regulations.

Attics, crawl spaces, vaults, manholes, concrete pier columns and elevator pits are all examples of confined spaces that are common on many construction sites. It is imperative that these spaces be identified and marked, and that all employees be trained to stay away from them unless they are a designated and trained entrant whose work requires entry. Confined spaces must be designated as permit-required or non-permit spaces. Entry Contractors must determine which of the three methods of entry described in this Chapter will be utilized to gain access to each permit-required confined space. Contractors will find that preplanning their work is essential to both making the work safer and working in a more productive manner. For the times where the hazards in confined spaces cannot be eliminated, the contractor will be required to develop and execute a full permit-entry plan that includes on-site supervision, dedicated attendants, stand-by rescue methods and specialized training. This procedure is discussed in the following chapter.

4 ENTRY WHEN HAZARDS CANNOT BE ELIMINATED

As discussed in previous chapters, it is always preferable to take steps to eliminate all hazards prior to entry. Doing so may allow the temporary reclassification of a permit-space due to the fact that the space has been rendered safe; however, it is not always feasible to eliminate all hazards, and the standard differentiates between control and elimination. Elimination of a hazard means that steps have been taken to remove the hazard from the space. For example, completely shutting down and locking out the energy to a piece of hazardous mechanical equipment located inside a confined space is an example of elimination of a hazard. If it were determined that this same piece of equipment needed to remain energized during entry, but the entrants would utilize personal protective equipment to guard themselves against the equipment, that hazard would be controlled, but not eliminated. This is similar in nature to using continuous mechanical ventilation to render the atmosphere in a space breathable; the hazard is being controlled by mechanical ventilation, but the potential has not been eliminated. Specifically, the OSHA standard defines control as using engineering methods to reduce the level of a hazard inside a confined space. This is an important distinction because in order for a permit-space to be reclassified to non-permit, all hazards must be eliminated as opposed to just controlled. When hazards cannot be eliminated, entry would need to be made following all of the rules outlined in the new regulations for permit-required confined space entry. There are circumstances on construction projects that may necessitate a full permit entry, including:

- Entry to perform work that is hazardous (welding or cutting for example)

- Weather conditions that expose an entrant to extreme heat or cold

- Atmospheric conditions that can't be controlled through mechanical ventilation and require supplied air or a self-contained breathing apparatus (SCBA)

- Internal conditions or configurations that could cause an entrant to become trapped or engulfed

- Entry into a space with energized and unguarded equipment

These circumstances, or entrance into any other permit-required space that contains a hazard that cannot be eliminated, would dictate entry following the full permitting process described in 29 CFR 1926.1205. The basic steps for entry into a permit-space are as follows:

1. The Entry Contractor must establish a written confined space entry program

2. A competent person for the Entry Employer must evaluate the space and establish the entry protocols and procedures that must be followed in order to control hazard exposure

3. An Entry Supervisor must be assigned to oversee the entry and make sure that all of the established entry procedures are followed

4. An Entry Attendant must be assigned to watch the entry into the space and to constantly monitor the entrants' health and safety while they are inside the space

5. A means of non-entry rescue must be established and all required equipment set up on-site at the entry into the space

6. If non-entry rescue is not feasible, a stand-by rescue team that is familiar with the space and the hazards it contains must be designated and available to respond immediately as needed to ensure the safety of the entrants

7. All of the designated individuals serving in these roles must receive training related to entry into the space, the hazard(s) that space contains and any PPE that will be utilized

8. All of this information, including the names of the individuals who will serve in all of these roles, must be recorded on a written Permit document that is posted at the entrance to the space

9. After the work is performed and entry is terminated, it must be documented on the permit and the permit must be closed out and retained

10. A post entry review must be conducted to evaluate the effectiveness of the established procedures and to identify any unanticipated conditions – this information must be shared with the general contractor (Controlling Contractor) and the owner (Host Employer)

Each Entry Employer must have their Competent Person evaluate the permit space to designate entry and exit points and to establish the procedures that must be completed prior to each entry. That Competent Person must identify each anticipated hazard and then establish procedures designed to either eliminate or control them. A best practice is to document these hazard evaluation and abatement procedures in a pre-task job hazard analysis (JHA), which can then be used as the basis for a training session with the individuals who will be involved in the actual entry procedure.

ASSIGNING AND TRAINING PERSONNEL

The roles that must be filled include that of the Entry Supervisor, Entry Attendant, Rescue Personnel and the Entrants. The basic duties of each role are similar to those defined under the confined spaces rules for General Industry, except the new rules for the Construction Industry do contain some differences in recognition of the fact that construction sites are typically active and constantly changing. The roles and responsibilities that must be assigned are as follows:

- Entry Supervisor - the qualified person responsible for determining if acceptable entry conditions are present at a permit space, for authorizing entry and overseeing entry operations, and for terminating entry and cancelling the Permit

- Entry Attendant - an individual stationed outside the permit space who verifies authorized entrants and monitors their health and safety, prevents entry by unauthorized personnel, orders evacuation when deemed necessary, performs non-entry rescue and activates the emergency action plan when needed – this individual can perform no other duties that might interfere with these responsibilities

- Authorized Entrant(s) – one or more individuals who will enter the space to conduct work and have been trained in the hazards that might be encountered, the procedures to be followed, the PPE required, and the rescue method and procedures that will be employed in case of an emergency

- Stand-by Rescue Personnel – a team that has been trained in the conditions and hazards that are present or that could occur within the permit space, is available to respond in a time frame that is appropriate given the hazards and potential injuries, is equipped and trained to successfully enter the space, extract injured entrants and provide first aid and CPR – rescue teams are required to practice actual rescue operations from the space or a representative mock-up of the space prior to attempting an actual rescue, and then at least once every 12 months

Under the rules for construction, the Entry Contractor has some discretion in determining how these roles are assigned. For example, the Entry Supervisor and the Entry Attendant can be the same person if the Entry Contractor determines that is appropriate under the site specific circumstances. In the case of a contractor making entry into a single space, it might make sense for these two roles to be assigned to a single qualified individual. An Entry Attendant may be assigned to watch over more than one space if it is determined that a single person can manage multiple spaces and entry crews. This may be appropriate in the case of simultaneous entry into two spaces that are in close proximity to each other, so that a single person can see both of them from one location. In this case, if something were to occur that required the complete attention of the attendant at one of the spaces (for example a non-entry rescue), entry into the other space would need to be immediately terminated.

Entry Attendants and Supervisors are prohibited from entering the permit space. Their duties are meant to be conducted outside the space, with their primary objective being the safety of the entrants. The Entry Attendant must have a positive means of communication with the entrants from outside the space. In some situations it may be appropriate to use voice only. If entry is being made into a 10-foot-deep pit that is completely open at the top, the attendant can probably see and speak to the entrants without the need for any equipment; however, entry into a more complex space, or a space where entrants will be out of sight, may require the use of two-way radios. Again, the advances in technology have resulted in the availability of very low cost and reliable two-way radios, making this an easy requirement to meet.

The Attendant must maintain their position outside the space in order to relay instructions and monitor the health of the entrants. They cannot leave their post for any reason while the Entrants are in the space unless they are replaced by another authorized Attendant who is listed on the Permit. It is their job to keep unauthorized personnel out, order evacuation of the space if unforeseen conditions appear, call for help if needed, or to affect a non-entry rescue. Rescues requiring entry into the space can only be conducted using trained rescue personnel.

Note that in the case of entry into a permit space by multiple contractors at the same time, or in the case where work is being performed by other trades on the site that could impact the entry operations, the Controlling Contractor is required to work with the Entry Contractor to coordinate operations. The best practice would be to limit permit-required space entry to one contractor at a time. The so-called stacking of trades is typically not a good idea out in the open, and it is rarely a good idea inside of a confined space. When conditions necessitate entry by multiple trades, proper preparation and coordination by the Controlling Contractor is critical to ensure the safety of the entrants. By stating that the Controlling Contractor must coordinate activities when multiple employers (i.e. more than one trade contractor) are entering a space simultaneously, 29 CRF 1926.1203(h)(4) has the effect of ensuring that the general contractor is actively involved in this process. They do not have the option of just leaving it to the subcontractors to work it out among themselves.

Chapter 2 discussed training for all personnel to educate them on the general hazards of confined spaces and to ensure that all construction employees understand the importance of staying out of confined spaces unless they have been trained and authorized to enter for the explicit purpose of conducting work under their contract. Personnel who are going to be involved in the actual entrance into a permit space require additional training. These personnel need to undergo training that discusses both the nature of the general hazards anticipated in this type of space, as well as the specific hazards and abatement solutions outlined by the Competent Person. Their training must also contain information on the mode, signs and symptoms that could result from any exposure; the method that will be used for communication, and the procedures that will be followed if any unanticipated conditions are encountered or a rescue needs to be undertaken. Using the JHA that was completed during the assessment of the space will help to ensure that the entry team has been trained in the specific safe entry procedures that were established during the assessment, as well as the PPE that has been specified and the rescue procedures that are feasible. All of this specific training must be documented and the Entry Employer is required to keep this documentation for each trained employee for as long as that person remains employed.

ESTABLISHING PROCEDURES AND COMPLETING THE PERMIT

After the safe entry procedures have been established and entrance personnel have been assigned and trained, this information must be transferred to the entry Permit. The Permit must include the names of the Entry Supervisor, the authorized Attendant(s) and the authorized Entrant(s). It must list all procedures and any steps that must be taken to eliminate or control hazards prior to entry, including a detailed list of any PPE that is required for entry. This should match the information established by the Competent Person in the JHA and may include things like removing and locking out hazardous energy, setting up guard rails, ventilating the space, testing atmospheric conditions and any other Permits that may be required for work inside the space, such as "hot work" Permits. The Permit must also list the means that will be used for rescue and identify the stand-by entry rescue team if a means of non-entry rescue is not being used.

Prior to entry, the entry supervisor must confirm that each of these steps have been completed and that the space meets the conditions for entry that have been established on the Permit. This includes verification of proper PPE and any testing that might be required. The Entrants must be given the opportunity to witness this verification, including witnessing any testing or monitoring that is performed, for example, atmospheric testing to ensure a non-hazardous atmosphere. Once the acceptable entry conditions are verified, the Entry Supervisor must sign the Permit to authorize the entry. At this point the Entry Attendant assumes responsibility for monitoring the entry personnel and entry may commence.

During entry, the Attendant stays posted at the entrance to keep unauthorized personnel out and monitor the health and safety of the Entrant(s). If any unanticipated hazard is encountered, or a prohibited condition arises during the entry (such as an unauthorized person entering the space), the regulations dictate that entry must be terminated, Entrants evacuated immediately, and the Entry Supervisor notified. The confined space rules for construction allow the Entry Supervisor to either cancel or suspend the Permit depending on the circumstances. If the conditions that required entry to be terminated are temporary and do not change the configuration of the space or create any new hazards, the Entry Supervisor may suspend the Permit and then allow re-entry under the same Permit. In this case, it is the Entry Supervisor's responsibility to assess the situation and verify that everything has been restored to permissible entry conditions before allowing re-entry. If entry was terminated due to a change in the configuration of the space or due to the discovery of a new or unanticipated hazard, the Permit must be canceled by the Entry Supervisor. Re-entry will require a new assessment by the Competent Person, potential additional training, and the issuance of a new Permit reflecting the updated conditions.

DON'T FORGET THE SAFETY DATA SHEETS!

If work inside of the permit-required space includes the use of hazardous chemicals, don't forget to have the product's Safety Data Sheet (SDS) available. A best practice would be to include these sheets as supporting documents that are attached to the Permit and made available on the site. The SDS will contain information about proper handling, required PPE, and first aid and emergency procedures. If a rescue is being conducted due to a mishap with a hazardous chemical, this is information that could prove critical during the rescue and treatment of an entrant. That is not the time to be trying to find the SDS in the binder that's kept in the trailer, and it's not the time to be trying to figure out how to get the electronic copy on your phone transferred to the paramedics who are rushing an exposed entrant to a hospital. The OSHA regulations require in §1926.1211(d) that an SDS be provided to the medical facility treating an exposed entrant, but keep in mind that when working in permit-required spaces with hazardous chemicals, the need goes beyond fulfilling the regulatory requirements.

STAND-BY RESCUE REQUIREMENTS

The OSHA regulations are clear when it comes to requirements for stand-by rescue when entering a permit-required confined space under this process. It is required and must be documented on the entry Permit. When the space is evaluated by the Competent Person for determining safe entry procedures, it must also be evaluated for rescue options. Where feasible, a means of non-entry rescue must be selected. This is commonly seen in the field as a tripod hoist set up over a manhole, or some other type of hoisting equipment erected at the entrance to a vault. Non-entry rescue is always preferable due to the risk of additional injuries and casualties to any rescuers trying to enter the space. In fact, there are, unfortunately, many documented examples of situations that resulted in not only the fatality of the original Entrant, but the additional fatality of untrained and improperly equipped rescuers who tried to enter a space to pull out an unconscious co-worker.

Non-entry rescue procedures should be carefully evaluated for feasibility given the actual conditions of the space. If it becomes apparent during an entry that the procedures that have been selected for non-entry rescue would not actually work if something went wrong, that is grounds for terminating the entry, cancelling the Permit and consulting with the Competent Person to reassess. The re-assessment may determine that non-entry rescue is infeasible and the work may require a stand-by rescue team who has been properly trained and equipped to enter the space to conduct a rescue. Before making this determination it is always advisable to explore all options, including things like cutting in an additional exit or resequencing the work to avoid hazards.

As stated earlier, the most common means for providing non-entry rescue is to provide a way to hoist or pull an Entrant out of the space. The Entry Attendant's responsibilities include operating that equipment from outside the space. This requires the Entry Attendant to be trained in the use of the rescue equipment and the complexities of pulling an unconscious person out of a confined space. All equipment must be set up on site prior to making entry and the Entrant(s) must be connected to the equipment at all times. The Attendant should also be trained in first aid procedures that match any anticipated injuries, and in how to activate emergency personnel. In the event of an emergency, the Attendant must have a means of activating the emergency action plan without leaving their post, and should be trained to activate the emergency plan immediately so that other personnel are reacting and mobilizing while the injured Entrant is being extracted from the space.

FIGURE 6 – MAKING PREPARATIONS FOR NON-ENTRY RESCUE

Where non-entry rescue has been deemed infeasible, the regulations require the assignment of a stand-by rescue team. The regulations allow some flexibility in determining what is needed in these situations, but the rule states that the capability and response time of the rescue team must match the hazards and potential injuries that could occur in the space. There are also several very specific requirements in the standard that make reliance on the local fire department an unlikely means for meeting the intent of this standard without having them involved in on-site planning, training and entry processes.

The stand-by rescue team is required to be trained not only in general confined space rescue, but must also be briefed in advance on the specific hazards that are anticipated in that particular space. These specific conditions may dictate additional training, equipment or PPE. The response time must be appropriate to the anticipated hazard. In other words, if the anticipated hazard is asphyxiation due to a toxic atmosphere, an average local response time of 5 minutes by the municipal fire department is not going to be adequate to save an Entrant. By the time they arrive, assess a situation they are seeing for the first time, and gear up, they will likely be performing a recovery, not a rescue. The stand-by rescue team must also be in communication with the entry team. The rescue team needs to know they are on stand-by and if they become unavailable they must immediately communicate that to the Entry Attendant so that entry can be terminated. This team must also be trained and equipped in the first aid procedures that are relevant to the type of injury that may occur in the space.

TRAINING RESCUE TEAMS

If something goes wrong and an entrant is injured or incapacitated during a permit-required confined space entry, they will rely on the designated rescue team to get them out of the space, provide first aid and get them transported for medical treatment. Regardless of whether the rescue will be a non-entry rescue or will require an entry rescue team, proper training is required. If the entry attendant is expected to hoist an incapacitated entrant out of a vault, they need to have practiced the procedure prior to executing the plan during an emergency. When you have an unconscious co-worker dangling from the end of a life line attached to their harness, it's not the time to be reading the instructions on how to operate the hoist. If the entrant is unconscious due to asphyxiation, the rescuer needs to be trained and prepared to perform CPR. If the work is being conducted in a remote area, or in a new residential community that does not have street signs or does not yet appear on the maps, it is going to be very difficult under the stress of dealing with an injured co-worker to explain to the first responders who to get to the work site.

These are all considerations that are critical to consider when training your entry and rescue teams and when developing a rescue plan. Non-entry rescuers must know how to activate emergency responders, they must know how to operate the extraction equipment, and they must be trained and equipped to provide the appropriate first aid and CPR while waiting for first responders to arrive on-site. Members of an entry rescue team must receive additional training that includes demonstrating proficiency as an authorized entrant along with the proper use of all PPE and equipment that they may be required to use during the rescue.

Rescuers should practice their designated rescue methods prior to being required to perform rescue duties and should practice at least once per year after their initial training. Practice should be done by means of simulated rescue operations in which rescuers remove dummies, manikins, or actual persons from the actual permit spaces or from representative permit spaces. Representative spaces must simulate the types of permit spaces from which the rescuers may be performing. This type of training is often conducted at a commercial training facility that has been specifically designed and constructed for confined space rescue training.

In many instances, particularly in remote areas, areas that do not have immediate access to a municipal technical rescue team, or conditions where the hazard in the space is an atmosphere that is immediately dangerous to life or health, the use of the local first responders, or using 911 as your rescue plan, is not appropriate. Consider the well-being of the Entrant(s) when making this determination.

For situations where the anticipated response time from a municipal first responder will be adequate to save an injured or incapacitated Entrant, this may be an acceptable component of an entry rescue plan, but it still requires communication and coordination. The requirements stated herein, including being briefed on the specific configuration and hazards of the space, being trained on entry into the space and being properly equipped, all still apply. This will require a pre-entry discussion with the first responders to ensure these criteria have been met. Keep in mind that not all municipalities have a technical rescue team (TRT), and even if they do, that TRT may not be specifically equipped to perform an extraction from your space. There must also be an established means for notifying the responders that entry operations are being conducted, and for the responders to notify the entry team in case they are made unavailable, such as would occur if they must respond to another incident.

If conditions rule out the use of a state or local first responder as your entry rescue team, the options are to contract with a private rescue company to have them available and, most likely, on-site during entry operations. A company can also choose to train its own personnel to become rescuers. Whether services are in-house or contracted out, the training requirements and capabilities of the rescue team are the same, so the decision may come down to economics and logistics. For example, a company that regularly engages in the business of coating and sealing tanks and vaults may find it advantageous to train and equip their own in-house rescue team(s), while a contractor who only rarely encounters a situation that calls for entry rescue may find it more appropriate to contract out for these services.

CLOSING OUT THE PERMIT

When entry is complete and the Attendant has verified that all Entrants are out of the space, the space is to be secured and the Permit is to be cancelled. If the Permit and procedures included any post-entry procedures to return the space to its original condition, those should be performed and verified by the Entry Supervisor who will sign off on the Permit stating that entry is complete. The entry team must document any conditions encountered in the space that were unanticipated, including anything that resulted in Permit termination or suspension. If entry went as anticipated and the JHA accurately addressed all hazards, this should be noted on the Permit.

The Permit is to be retained by the Entry Employer for a period of at least 1 year. The Entry Employer must conduct a post-entry review session to evaluate whether or not their procedures are working to keep their employees safe. This review may be done annually using all completed Permits, but it should be documented and the entry team employees should participate and be encouraged to provide input. In addition to the Entry Employer review, the documentation on the entry must be forwarded to the Controlling Contractor on the site so that they have information that may be pertinent to other trades entering the space throughout the project cycle. The Controlling Contractor is responsible for forwarding information on all permit-required confined spaces and actual entry information to the Host Employer (Owner) so that they can incorporate that information into any future maintenance entries during the life of the structure.

INFORMATION THE OWNER NEEDS

One of the primary goals of the OSHA rules for confined spaces in construction is to facilitate communication that we know has not readily flowed in the past. For example, on a project I was involved with, the entry team disturbed a previously unseen nest of scorpions located inside of a floor drain at the bottom of a deep pit. This information needs to be transmitted from the contractor back to the owner so that the owner can communicate this potential biological hazard to other construction and maintenance crews that are sure to enter the same space in the future.

CONCLUSIONS

The rules for confined spaces in construction are an extensive set of regulations that were enacted as a response to continuing fatalities in these spaces that include not only the workers themselves, but the would-be rescuers who rush in to save their co-workers without proper training or equipment. The rules allow the reclassification of a permit-required confined space if all hazards can be eliminated prior to entry, and they allow a modified permit-entry procedure when all hazards are eliminated except for a potential hazardous atmosphere that is controlled by continuous mechanical ventilation. This modified procedure eliminates the OSHA requirements for dedicated Entry Attendants and stand-by rescue. These options can greatly reduce both the cost of entry and the risk of injury to the entrants and should be evaluated and considered before undertaking the considerable additional steps that are required for a full Permit entry.

When a full Permit entry is required to complete work, the Entry Employer must have a Competent Person completely asses the space and design the steps and procedures that will be utilized to keep the Entrants safe. The Entry Supervisor is responsible on-site for ensuring that all steps are followed and that the entry team personnel understand the hazards and have received proper training. A dedicated Entry Attendant is responsible for remaining outside the space at all times to monitor the entry and the health of the Entrants. The Attendant must also have the means and the training to activate the emergency action plan and perform a non-entry rescue should the need arise. In any case where non-entry rescue is infeasible, there must be a stand-by rescue team designated that is trained and ready to enter the space and rescue an entrant if something goes wrong.

The new rules found in 29 CFR 1926 Subpart AA are expected to save lives and prevent injuries if they are properly enacted by contractors who's work requires them to enter confined spaces. The rules regarding posting, communication and hazard awareness training for all personnel who may work near confined spaces or on sites where confined spaces exist are meant to keep everyone on a construction site safer by ensuring that they stay out of these spaces unless they are trained, equipped and authorized for entry to conduct work.

5 JURISDICTIONS, AUTHORITIES AND EXCEPTIONS

Federal OSHA's jurisdiction generally does not extend to State and Local government employees; however, a total of 28 states and territories operate their own OSHA-approved state plans which do extend to State and Local government employees. All of these state run plans are required to adopt and enforce rules and regulations that are "at least as effective as" the federal regulations, and in fact they typically adopt the federal standards as-is, with no changes. On those rare occasions where a State plan adopts a rule that is not identical to the federal standard, the changes will generally result in more stringent regulations. In fact, these states are prohibited from adopting rules that federal OSHA would consider less effective.

One result of these state plan programs is that state and local governmental employers fall under the same rules and regulations as private employers, and the states' occupational safety and health departments could consider a government agency either a Host Employer or a Controlling Entity under the confined spaces standard. In addition, an agency that sends an employee, for example an inspector, into a permit-required confined space would be considered an Entry Contractor. This is an important consideration as there are no exemptions for inspections under this standard. To be clear, if an inspector must enter a permit-required confined space, that individual and their employer are subject to the same entry requirements imposed on any contractor doing work in the space, even if that inspector works for a government agency.

The practical implications here are that contractors, owners and agencies need to start thinking about working together and coordinating entry, or better yet, considering whether or not owner or agency entry is even really necessary. When a space can be rendered safe and temporarily reclassified, entry by an inspector or owner's representative may be justified, but as the hazards increase the question should be posed to all parties involved... "is sending one more person into this space really in the best interest of the owner, agency, etc.". The argument can certainly be made that the technology probably exists in today's industry that would allow for things like remote viewing of the conditions inside of a space. It is very easy to take digital pictures and even video from inside a space at the conclusion of a task so that it can be viewed by inspecting parties later. It is even feasible, and not that complicated, to have the contractor's Entrant send live video feeds out of the space for viewing, resulting in fewer people needing to enter the space and be exposed to hazards.

NO EXCEPTIONS FOR INSPECTIONS OR SHORT DURATION WORK

Some OSHA standards allow exceptions for inspections or even for very short duration work were the contractor can prove a greater hazard may exist if they attempt to follow the requirements. For example, in fall protection, an exemption exists that would allow an inspector to look over completed work on a roof without tying off if it is safe to do so. Another example might include walking a short distance on a flat roof to install the guard rails. The confined spaces standard does NOT contain any of these exceptions. An inspector entering a permit-required confined space would be subjected to the same hazards as any other entrant, and they are there for subjected to all the same requirements for entry. Likewise, an entrant is not exempt from the rules just because they are entering quickly to get a photo or a part number; in fact, the rule clearly states that entry to complete work to render the space safe must be done using the permit entry procedures.

When the situation does call for entry by an inspector or agency representative, all parties need to understand the risks and responsibilities. In general safety terms, an agency that sends an employee (an inspector in this example) out to any ongoing construction site is considered an exposing employer under the multi-employer worksite rules that are generally referred to as the "Rules of Construction". Even though the work site is under the control of a general contractor and hazards are being created by subcontractors, any employer sending an employee to work in the proximity of these hazards is considered an exposing employer, and in the 28 states and territories that operate a state plan, government agencies do not get a pass. This should be considered a good thing, i.e. government employees get the same protection and right to a safe work environment as their private sector counterparts; however, these requirements are sometimes lost on some agencies.

Under the confined spaces rules, even agencies that are not included in the 28 states and territories who fall under a state OSHA plan need to understand that Controlling Contractors and Entry Contractors still have a responsibility to protect anyone that enters the space under their watch. The case can certainly be made for a contractor denying entry to an inspector who is not trained or equipped to handle the anticipated hazards. Again, collaborating on an alternate means of inspection or verification that keeps them out of the space may be in the best interest of all parties.

There are 26 states and 2 U.S. territories that have their own OSHA-approved occupational safety and health programs called state plans. The following 22 states or territories operate state plans that cover both private sector and state and local government employers: Alaska, Arizona, California, Hawaii, Indiana, Iowa, Kentucky, Maryland, Michigan, Minnesota, Nevada, New Mexico, North Carolina, Oregon, Puerto Rico, South Carolina, Tennessee, Utah, Vermont, Virginia, Washington and Wyoming. Five additional states and one U.S. territory operate state plans that cover public sector workers only: Connecticut, Illinois, Maine, New Jersey, New York, and the Virgin Islands.

From time to time there are certainly going to be projects where an agency representative will be present during a confined space entry. I have several projects where construction is being done inside of an existing state owned facility that contains many confined spaces. In this case, the agency has trained and highly qualified personnel who are probably more familiar with these spaces than the contractors. The agency has already identified the spaces as permit-required, posted notices and secured them from entry. When work requires entry by an agency employee together with a contractor's employee, it is a relatively simple matter of coordination. The agency already has a confined space plan and permit entry procedure, along with trained and qualified Entrants. The general contractor can simply act to facilitate communication and concurrent entry by the two parties.

Another example might be a scheduled bridge or structural inspection that involves several parties entering culverts or box girders for extensive structural investigations. This is the example that is going to take some additional planning that may or may not be occurring out there in the field today. The first thing that needs to happen is the inspection and evaluation by a Competent Person. Ideally this would already be done by the agency that owns the facility. I make this statement out of comparison to any other privately owned facility subject to regular maintenance where the facility owner would have been expected (under the

General Industry Standards) to have identified, evaluated, marked and secured its permit-required confined spaces. If this has not been done, and you are a contractor whose work requires your employees to enter what is clearly a confined space, you must have your Competent Person evaluate and classify the space.

In the example of a box girder that must be entered and inspected, a Competent Person may evaluate the space and determine (1) that it fits the definition of a confined space and that (2) it contains the potential for a hazardous atmosphere in the form of carbon monoxide accumulation inside the space. In this case the two parties (the contractor and the agency) are going to have to work together to form an entry plan and procedure. Again, referring back to the levels of entry described in this book, it is always preferable to work towards eliminating as many of the hazards in the space as possible, and then controlling what is left. In this example, we will assume that all hazards are eliminated except for the potential for a hazardous atmosphere. The recommended procedure at this point would most likely be to test the atmosphere as it currently exists in the space to determine a starting point. This would of course best be tested without any person entering the space, i.e. drop in a sampling pump or attach the meter to a pole and push it into the space without making entry. These initial readings may help the competent person determine the safest plan for entry. It may be safe to enter as is, or it may need mechanical ventilation and re-testing. Either way, you are ideally working towards utilizing the alternate procedure for entry (described in Chapter 3) when all hazards are eliminated except for the potential of a hazardous atmosphere. This means that continuous ventilation and monitoring will be required during entry, but you may determine that you can eliminate the need for Attendants and stand-by rescue. Even so, any individual who will enter the space needs to be properly trained and briefed on the potential hazards, and this needs to be a formal, documented process.

Experience shows that there are some facilities that do receive this level of attention such as utility vaults where it is not uncommon to see them properly marked and entry by the agency's maintenance crews is commonly made under a formal permit entry plan. However, there is evidence of many more that are unmarked and not commonly treated as confined spaces, or at least not as permit-required spaces. Probably the most prolific example of this is sewer and storm drain manholes, and this is where these new rules will need some serious attention. Is a typical manhole a confined space? Absolutely…its big enough to enter, not designed for continuous occupancy and it has limited means of access. But is a manhole a permit-required space? Probably…certainly any active sewer manhole has not only the potential, but the likelihood of a hazardous atmosphere, along with the potential for engulfment depending on the configuration. What about a storm drain manhole? Again, probably…it would be very difficult to justify a statement that an active storm drain manhole did not contain the potential for a hazardous atmosphere.

These are the types of situations where a contractor is probably going to have to take the lead, and this starts with an assessment by their Competent Person. If the space contains hazards or has the potential to contain a hazardous atmosphere, it should be classified as a permit-required space. In the assessment, the Competent Person should ideally be working towards eliminating all hazards in order to temporarily reclassify the space to non-permit. Remember, this needs to be a formal process and it needs to be documented. All Entrants should review this documentation so they understand what hazards existed and how they were abated.

If the Competent Person determines that a space has the potential for a hazardous atmosphere, it will be difficult to eliminate that potential. Remember, if it has the potential, it is a permit-required space, even after you ventilate it and get acceptable atmospheric readings.

Pulling this all together and applying it to the example of the box girder inspection, once the Competent Person evaluates the space and determines that the only remaining hazard is a potential hazardous atmosphere, a formal plan can easily be developed that may be as simple as providing continuous mechanical ventilation and continuous atmospheric monitoring. The Entrants from the agency and the contractor can review the plan with the Competent Person and make entry under the modified confined space entry procedures described herein. This does not need to be a long drawn out process, but it does need to be a formal documented process. It is not okay to simply arrive at the structure, open the access covers and let it sit for a while to "air out". The rules clearly require mechanical ventilation and continuous monitoring, along with training for the Entrants so they all understand what to do if they encounter things like equipment failures, atmospheric meter alarms, or other unanticipated hazards.

This can all be done efficiently and even on short notice when the appropriate personnel are available, but there will need to be discussion as to which party takes the lead; the agency or the contractor. In my first example of performing construction inside the State controlled facility, they would most definitely take the lead and expect the entry contractor to show compliance. In the example of entrance into something like a heavy civil structure where the agency may not have previously taken any steps towards evaluation or classification, it is very possible that the contractor will end up taking the lead and insisting on compliance from the agency and its Entrants. To some, the contractor taking the lead and insisting on agency compliance may seem counterintuitive, but it is important to keep the situation in perspective. If you control the space and the entry, you put lives at risk if you allow entry by untrained and unprepared individuals. This includes the lives of the untrained entrants and all other entrants inside the space, as well as the lives of any potential rescuers if this becomes necessary.

Chapter 6 discusses the differences between the OSHA General Industry Standard (§1910.240) and the Construction Industry Standard (§1926.1200), and it can be seen that in many cases a plan that complies with one set of rules may comply with both sets of rules, but the question does often arise… "Which standard is to be followed if my company is doing construction work inside of a facility that is regulated by the General Industry rules?" or "What standard is to be followed if my company is doing construction AND general industry work in confined spaces?" OSHA has answered these questions in several documents including its website and the Federal Register notice where the standard was originally published. They clearly state that "An employer whose workers are engaged in both construction and general industry work in confined spaces will meet OSHA requirements if that employer meets the requirements of 29 CFR 1926 Subpart AA - Confined Spaces in Construction."

There are a few exceptions to these rules that exist generally because another standard already exists to protect workers who enter a specific type of confined space. In the OSHA Standard itself it states that activities excluded from the standard are as follows:

- Diving - regulated by 29 CFR Part 1926 subpart Y

- Excavations - regulated by 29 CFR Part 1926 subpart P

- Underground Construction, Caissons, Cofferdams and Compressed Air – regulated by 29 CFR Part 1926 subpart S

In addition to these expressly excluded activities, OSHA also entered into a settlement and issued an enforcement policy in April 2016 stating that any work on or directly related to telecommunications lines and equipment by telecommunications employees in an existing telecommunications manhole or vault will not be subject to citation by OSHA under the Confined Spaces in Construction standards, as long as such work is performed in accordance with the Telecommunications standard, 29 CFR 1910.268, and the hazards associated with such work are addressed by that standard.

Put in practical terms, work being done that is already governed by one of these four existing OSHA Standards (Diving, Excavation, Underground Construction and Telecommunication) comes with its own specific set of rules. Contractors are cautioned however, that scopes of work can overlap and invoke a need to follow both sets of rules. For example, a contractor installing storm drains would follow the excavation safety standards (§1926 Subpart P) during the trenching and installation of pipe; however, if they have a worker enter one of the newly constructed manholes to perform grouting of the riser segments they would have to comply with the confined spaces rules. Or, if a telecommunications worker must enter a vault with an atmosphere that cannot be made safe before entry as required by the Telecommunications Standard, that entry would need to comply with the confined spaces rules. Similarly, work installing new telecommunications manholes or vaults and work to install new ducts between existing manholes or vaults is not considered to fall under the scope of the telecommunications standard.

Finally, it is important to remember that any work being performed inside a confined space may also have its own set of specific rules that must be followed in addition to the confined spaces rules. One example of this would be welding and cutting. Not only does a contractor need to comply with the confined spaces rules for entry, but it must also follow the rules contained in §1926 Subpart J for welding and cutting. Some additional examples include any work that must comply with the requirements for electrical hazards (§1926 Subpart K), hazardous waste operations requirements (§1926.65) or chemical process management requirements (§1926.64).

6 COMPARISON TO OTHER STANDARDS AND REGULATIONS

It was argued by some during the OSHA rule making process that the construction industry did not need its own set of rules and it was suggested that the industry just be mandated to adhere to the existing General Industry standards contained in 29 CFR 1910.146. The argument countering this position was that construction sites are different than general industry facilities because they are constantly changing and evolving and they almost always have multiple employers at the project. Treating confined spaces in construction differently than fixed general industry sites makes sense in order to address the issues that arise at multi-employer sites, and to assign specific responsibilities to the various parties that can be involved on a construction project. For the most part, things like terms and definitions have been carried over from the General Industry Standard, and the general concepts that work to mandate safe entry conditions are the same; however, the Construction Industry Standard does contain some differences to address construction conditions. OSHA also took the opportunity to make some updates that are probably justified by technological advances and from lessons learned in general industry.

WHAT'S DIFFERENT FROM THE GENERAL INDUSTRY REGULATIONS?

Since many construction projects may begin with a site that contains unidentified spaces, or in which confined spaces will not even be constructed until later in the project, the Construction Industry standard requires a Competent Person to evaluate the work site and identify all existing and future confined spaces. It is important to note that there is also nothing in the new regulation or any OSHA published literature that suggests the competent person evaluation may be eliminated in an existing facility which has already identified its permit-required spaces. In effect, this evaluation is a new mandate that effects every construction site to one degree or another. The evaluation may be simple on some sites, it may include documentation of existing permit spaces previously identified at an existing facility, or it may require ongoing or periodic re-evaluation as a project progresses and physically takes shape. In any case, the evaluation must be performed.

One of the single most significant differences between the construction rules and the general industry rules involves the inclusion of more detailed provisions requiring coordinated activities at the multi-employer worksites which are the norm in the construction industry. The specific assignment of roles and responsibilities is meant to mandate a flow of information back and forth between parties on the site to ensure that hazards are not introduced into a confined space by workers performing tasks outside the space. Because the construction regulations add some definitions and take steps to assign specific responsibilities, it is extremely important that all parties understand their assigned roles and the accompanying responsibilities as outlined and discussed in Chapter 2.

Contractors who control construction projects are best advised to leverage these new mandates to gather information to help them determine the best way to construct a project. If forcing an evaluation and discussion of responsibilities prior to beginning work on-site results in a re-sequencing of work that keeps employees out of hazardous confined spaces, the additional benefits could be increased productivity and improved quality due to work being completed before a space is enclosed.

In another recognition of the ever changing conditions that can exist on a construction site, the construction rules also clarify the ability to suspend a Permit rather than cancel it when conditions inside of a space do change due to activities by workers performing tasks outside the confined space. In cases such as these, entry may resume under the suspended Permit once conditions are returned to the entry conditions listed on the Permit.

Additional variations include:

- Requiring continuous monitoring of atmospheric hazards whenever possible instead of periodic testing – this is facilitated by significant technological advances in the equipment used to conduct the monitoring (see Appendix A)

- Requiring the continuous monitoring of engulfment hazards – This may apply in situations such as working inside of a storm drain catch basin where a potential for upstream flooding could result in engulfment of the entrants

- A clarification that requires employers who choose to rely on local public emergency services as a part of their entry rescue plan to get verification of the rescuer's abilities and training and to remain in contact with them to ensure they are notified when the public services become unavailable (when responding to another emergency for example) – See Chapter 4 for additional information

- Requiring employers to provide training in a language and vocabulary that can be understood by their workers – this is somewhat redundant since these general training requirements are already mandated elsewhere

Finally, the construction regulations include several clarifications of the differences between controlling hazards versus eliminating them. Isolation of hazardous energy through a lockout/tagout procedure is considered elimination of a hazard, however, since the 1926 Construction Industry standards do not contain specific lockout/tagout procedures and requirements, these basic procedures are described with some mandates within the confined spaces in construction regulations.

COMPARISON TO NFPA 350

About the same time OSHA was publishing these rules for confined spaces in construction, the National Fire Protection Association (NFPA) published a standard titled, NFPA 350: Guide for Safe Confined Space Entry and Work. NFPA is an ANSI Standards Developer and this document was approved as an American National Standard in December of 2015. NFPA publishes a number of guide documents that are utilized by some in the construction industry to develop best practices to keep their workers safe. NFPA 70E, for example, is an excellent guide to working on electrical circuits that is used by many electrical contractors to develop rules for ensuring regulatory compliance and safe working conditions. This 2016 edition of the NFPA confined spaces guide is the first document published by this group that relates solely to work (other than rescue) in confined spaces other than the maritime industry, and it is an excellent guide for safety professionals looking for a detailed resource to help them develop policies and procedures.

In the preface to the NFPA 350 Standard, the developing committee acknowledges reviewing the existing OSHA General Industry Regulations, but they do not acknowledge the much newer (about 20 years newer) Construction Industry Regulations. This is not surprising since a look at the development timeline for the NFPA document seems to put its development at about the same time as the new OSHA regulations. The preface discusses what the NFPA committee considered to be gaps in the OSHA regulation and notes a few of the major items in the NFPA Standard that go above and beyond the OSHA General Industry Regulations. One of the themes consistent throughout many of the NFPA standards, particularly those used outside of the emergency services industry, is the idea that the NFPA documents are meant to translate regulations into practice and that they are meant to go beyond the minimum requirements set by existing regulations and codes. This is immediately evident when reviewing the NFPA document.

This document mainly focuses on safe entry and work inside of a confined space. It does not contain any of the key language from the OSHA Construction Industry Regulations related to the assignment of roles and responsibilities (to host employers and controlling entities) and the mandated collaboration and flow of information. What it does do very well is to discuss best practices for qualifying an entry team, testing and evaluating confined spaces, monitoring atmospheric conditions, ventilating spaces, and developing an effective rescue plan. Using the NFPA 350 Standard to develop safe entry rules and procedures should result in compliance with the OSHA regulations pertaining to permit-required confined space entry in both general industry and on construction projects. For compliance with the entire OSHA Construction Industry Regulation, a contractor will also need to develop the steps described in this book for evaluating, collaborating and communicating prior to making the decision to enter a confined space.

As stated, this publication is a very complete reference guide to safe confined space entry, and it does go well above and beyond the minimum regulatory requirements in order to ensure the safety of workers in confined spaces. For example, this standard defaults to atmospheric testing of all confined spaces and does away with the language regarding testing spaces only if they are perceived to have the potential for a hazardous atmosphere. In other words, this standard takes the position that all confined spaces have a potential for a hazardous atmosphere and states that they should always be tested and monitored. The foreword to the NFPA Standard argues this is a small step toward ensuring that a space contains air that is safe to breath. Given the relatively small expense and quick and simple methods available today for testing (see Appendix A), this is an argument that many in the safety industry will find easy to stand behind.

The NFPA Standard also introduces several new roles for members of the entry team, including a *qualified gas tester* and *qualified ventilation specialist*. Again, given the lack of proper procedures that can be witnessed in the industry, these are probably valid positions to develop if your company regularly enters confined spaces. Improper ventilation of a space due to a lack of understanding of the physics behind air flow and ventilation can result in areas in a confined space that are left hazardous to the Entrants, and Appendix A in this book discusses some of the issues the construction industry faces related to the proper calibration and use of atmospheric testing equipment. These are the types of conditions that NFPA 350 tends to focus on and provide solutions for.

Another big area addressed in detail in NFPA 350 is the development, staffing and equipping of the stand-by rescue crew. While the OSHA regulations mandate non-entry rescue where possible and stand-by rescue teams where it is not, the OSHA requirements simply mandate matching the team's capabilities and response to the potential hazard. The NFPA document tells you how to actually do that. It includes a chart and discussion on types of hazards, number of Entrants and means of egress, and converts that information into the needed capabilities for the stand-by rescue crew. It also details the training and equipment they need to be effective. This information provides an informative (i.e. eye opening) look at what exactly you need to consider when deciding on your options for stand-by rescue.

NFPA 350 is a very complete reference to developing safe practices that will comply with OSHA regulations relating to entering permit-required confined spaces. Companies whose trades requires them to perform work inside of permit-required confined spaces will find this a valuable resource in the development of their entry procedures.

FIGURE 7 – PROPER VENTILATION METHODS ARE CRITICAL

COMPLIANCE WITH CANADIAN LAWS AND STANDARDS

For the most part, compliance with the OSHA Standard and the best practices described in this book will result in compliance with most Canadian laws and local jurisdictional requirements. In Canada, there are local jurisdictions which may have standards that differ slightly, and Canadian labor laws differ from the U.S. in that there are many more provisions and requirements for labor/management committees that may result in additional regulatory requirements. Using the Canada Occupational Health and Safety Regulations (SOR/86-304) which were current as of the publication of this book as a basis for comparison, it appears that regulations in both countries take the same basic approach to mandating safe work practices when entering confined spaces.

Section 11.1 of the Canadian Code defines confined spaces in a similar fashion as its U.S. counterpart, and requires written assessments of such spaces by a qualified person to determine the hazards and develop steps to mitigate them. It requires the assessment to be in writing and signed and dated by the qualified person making the assessment. It also notes that the assessment of existing, fixed spaces must be reviewed every three years (§11.2 (4)), and many of the documents related to confined spaces, i.e. the written assessments, entry plans and Permit forms, are required to be kept by the employer for a period of 10 years (§11.12).

The biggest difference between the U.S. and Canadian Standards appears to be the fact that the Canadian rules do not specifically differentiate between a permit-required space and non-permit space. The rules essentially treat all confined spaces in the same manner and require that they be assessed by a qualified person and that entry procedures based on that assessment then be developed as a part of a written entry permit program. In other words, entry into any confined space will require a written plan and entry Permit in order to ensure the safety of the Entrants (§11.3(a)). The degree to which that written Permit system requires specific steps such as atmospheric monitoring, mechanical ventilation, stand-by rescue teams, etc. will depend on the hazard assessment and the requirements generally line up with the U.S. rules.

The rules requiring emergency procedures and equipment (i.e. stand-by rescue) are found in §11.5 and again are similar to the requirements that would be spelled out for a standard permit-required entry under the U.S. Standard. §11.5 (1) (c) requires a trained Attendant to be posted outside the space at all times and the remainder of this section of the Code goes on to include specific requirements for rescue training, equipment and procedures. The Canadian Code is less prescriptive in nature and does not necessarily distinguish between non-entry rescue and entry rescue requirements other than to encourage having a means of non-entry rescue available and requiring special training and equipment when entry rescue is the planned means to be used during emergencies. One notable difference however is that §11.5. (1) (e) does specifically require that two or more people are in the immediate vicinity of the confined space to assist in the event of an accident or other emergency. It also requires that at least one of those additional individuals be trained in the established emergency procedures, be provided with the needed emergency equipment, and must hold a basic first aid certificate. Again, this is not unlike the U.S. requirements for a stand-by rescue crew except that the Canadian Code is very specific about the rescue personnel being present in the immediate vicinity.

Section 11.10 of the Code includes requirements for equipment to be used when mechanical ventilation is required and notes that this equipment must be equipped with an alarm to alert the Entrants in case the equipment fails, but also states that the alarm is not required if the equipment is continuously monitored by an employee who is in communication with the Entrants (such as the Attendant).

Another significant item to note is that Canadian laws and regulations do contain some rules pertaining to equipment inspections and certifications that may differ from their U.S. counterparts. One such difference relates to gas monitors used for atmospheric testing and this is noted and discussed in Appendix A. It is always advisable to ensure that any equipment being used within Canadian jurisdictions contains a CAN approval where one exists. This certification may differ slightly from an applicable standard commonly accepted in the U.S. such as a UL (Underwriters Laboratories) certification.

CONCLUSIONS

The OSHA Standard for Confined Spaces in Construction is (at the time of this publication) one of the newest standards related to entering and working in confined spaces and is generally based on the latest knowledge and technology. It is a very prescriptive standard, taking the approach of mandating specific steps and procedures to be followed. It is also a very complete standard that does a good job at imposing requirements aimed at making entry into a confined space a safer activity, and is the only national standard written specifically for the construction industry. Because of its prescriptive nature and its completeness, compliance with this Standard generally covers the main points of the other confined space entry rules that may be encountered; however, always review any requirements that may be imposed by local jurisdictions, by owners or by contract to ensure that you are in-compliance, and consider best practice documents such as NFPA 350 to aid in the development of entry procedures if you must work in permit-required confined spaces.

7 RESIDENTIAL CONSTRUCTION

In the past, there have been a few examples of OSHA regulations being applied differently on residential construction projects. The most notable being the fall protection standard found in §1926 Subpart M. When this regulation was first issued, there was also an enforcement directive issued which, for the most part, acknowledged that compliance with the rules on these types of projects was generally considered to be infeasible, and therefor directed contractors towards alternate means of providing safe working conditions. Since then, the technical means and methods have been developed and that temporary enforcement policy has been rescinded, removing one of the last examples of differentiation between residential and commercial construction when it comes to the application of safety standards. This theme has been carried forward under the rules for confined spaces.

When the Standard was first published, OSHA also published compliance assistance materials that included an OSHA Fact Sheet titled "Confined Spaces – Attics and Crawl Spaces" which was directed at the residential construction industry. Concurrent with a Petition for Review filed by the National Association of Home Builders (NAHB), that Fact Sheet was soon thereafter removed from the OSHA web site, however there are still State OSHA agencies that have similar fact sheets published and available (for example, Oregon OSHA's Fact Sheet OR-OSHA (2/16) FS-65). During review and discussion between OSHA and various entities representing the home building industry, OSHA did issue a temporary stay of enforcement on residential construction projects, but that directive was issued specifically to give the industry more time to comply with the new rules, as long as a company was able to show a good faith effort towards ongoing compliance efforts.

That temporary order has since expired and there is no differentiation between residential and commercial construction when it comes to enforcement of this Standard. This means that residential builders and their trade contractors need to be familiar with these rules and understand that all of the requirements discussed herein apply equally to them. In fact, even NAHB's own published fact sheet acknowledges the existence of confined spaces on residential construction projects:

> *Examples of confined spaces in home building may include, but are not limited to: manholes, sewer systems, stormwater drains, water mains, crawl spaces, attics, heating, ventilation, and air-conditioning (HVAC) ducts, and pits. (Published 5-11-15 by the National Association of Home Builders)*

In addition, other related industry groups, such as RESNET, the Residential Energy Services Network, have also issued opinions and guidance documents to their members relating to the confined spaces in construction standard. In an interpretation obtained by their legal counsel, RESNET provided this summary to its members:

> *The takeaway for RESNET raters is that if they are employing workers in crawl spaces or attics that qualify as confined spaces, they are going to have to comply with the standard. Sole proprietors who are themselves the only employee of the business are not subject to the OSHA standards. Employer-raters will need to comply with the evaluation process for identifying confined spaces and ensure that appropriate steps are taken for worker training prior to jobs, alleviation of hazards, and implementation of a permitting process (including attendants and rescue program plans) where applicable. However, where a site includes multiple employers, there is the ability to coordinate compliance, perhaps having the largest employer take responsibility for many aspects of the compliance.*

It is interesting to note that this legal opinion indicates in several places that compliance with the Standard is not required from a regulatory standpoint if Entrants are sole proprietors. While this is certainly a correct legal interpretation given that OSHA's jurisdiction and enforcement activities occur within the employer-employee relationship and interactions, the guidance seems to miss the point that the rules have been enacted to keep the Entrants safe, not simply as a regulatory exercise. All Entrants, whether they are employees who are granted specific protections under the OSH Act of 1970, or sole proprietors who are on their own to protect themselves, would be well advised to adopt compliance with the Standard as a means of best practice to assess spaces, identify hazards and prevent injuries and fatalities when entering confined spaces.

Before potentially over-reacting to a new set of regulations, it is important for residential contractors to keep in mind the distinction between confined spaces and permit-required confined spaces. For example, while crawl spaces and attics will typically fit the definition of a confined space, they generally do not have hazards that turn them into a permit-required space. Simply being a confined space does not by itself trigger additional requirements. It's the presence of dangerous hazards that turn the space into a permit-required space that necessitates the need to formally and methodically address those hazards before entry. The rules do mandate an assessment by a Competent Person to determine the difference, and conditions do vary. Normal service conditions in a space can also change when there is damage or other conditions that require repair. The Oregon OSHA fact sheet contains the following examples:

> *For example, a crawlspace may normally be dry, but if an employee opens the hatch and finds that the sewer line has broken and is leaking raw sewage, then it is a permit space until the problem is fixed and the sewage is removed. The nature of the work can also change the nature of the space. For example, if work within the crawlspace includes opening up the sewer line, the space is a permit space for the duration of the work. Once the pipe is closed and the space returns to its normal state, it can be re-evaluated as a confined space. Attics are typically confined spaces, but environmental conditions, such as heat during the summer, can change it to a permit space. The nature of the work can also change the nature of the space. For example, if the guards for electrical components need to be removed, exposing workers to live parts, it becomes a permit space. When the guards are replaced, the space can be re-evaluated as a confined space.*

Again, preplanning and giving consideration to the sequence of work in new construction on a residential project can be used to re-order work to make the site safer and facilitate compliance. Any resequencing of work that facilitates not having to put a worker into a confined space can make the project safer. For example, completing all work inside of the space that will eventually become a crawl space before constructing what will become the lid to that space results in eliminating the need for anyone to work inside after it gets turned into a confined space. This should be both safer and more productive. Likewise, locating hazardous equipment outside of confined spaces could result in a confined space avoiding the application of the permit-required label.

This concept of addressing construction hazards during the design phase of a project will be discussed further in Chapter 9 – Designing for Safety. The ability to evaluate the schedule and re-order the work on a project must be addressed by the home builder, who will typically fit the definition of the Controlling Contractor on a residential construction project. The home builders' participation in coordinating the schedule and the efforts of all their trade contractors is critical to ensure the safety of all workers on these sites. The recognition by the trade contractors of the types of confined spaces found on residential construction sites and the hazards they may contain is critical to ensure that their employees receive proper training and gain an understanding of the conditions that need to be avoided on these sites.

ADDITIONAL CONSIDERATIONS

It is worth including some additional discussion specifically related to residential construction and the type of confined spaces that may be seen on a residential construction project. First, it is important to remember that each space should be evaluated based on its attributes, and avoid generalities that may impose needless precautions. For example, the question that comes up quite often when discussing this topic is "Is an attic always a confined space?" The simple answer is "No", there are many instances in which an attic would not meet the definition of a confined space. In some areas, the term attic is used to describe a space (finished or unfinished) formed by traditional roof framing methods that is not considered a room, but is accessible by stairs and standard doorway. In this case the so-called attic would probably not fit the definition of limited means of access which is required for the space to be considered a confined space. In other areas, the term attic is used to refer to more of a crawl space that is formed by prefabricated roof trusses. In this case, the finished space is typically accessed through a hatch which is reached using a ladder. While this probably meets the definition of a confined space, it's important to remember that if the space is not yet enclosed by drywall, it may not fit the definition of limited access.

Remember too, just because the space fits the definition of a confined space, doesn't automatically trigger the restrictions assigned to permit space entry. The space must fit the definition of a confined space, and it must contain some other additional hazard that would trigger the permit-required designation. This could include a physical hazard, such as exposed and energized conductors, or an environmental hazard such as a leaking gas line or an extremely hot environment. The key is to determine whether the space contains a condition that would be immediately dangerous to life or health, or if a condition exists that would impede a person's ability to exit the space safely without assistance. If such conditions do exist, the best course of action is to take steps to eliminate or isolate the hazard prior to entry so that the space can be treated as a non-permit space. Not only would this result in a safer entry, it will most likely be more productive. Steps that can be taken to isolate hazards may include de-energizing and locking out electrical or mechanical circuits. Steps that may be taken to eliminate a hazard may include re-ordering the work so that the tasks that require confined space entry are performed prior to introducing any of the hazards that would render the confined space a permit-required space.

Additional considerations on residential construction sites involve the timing of the required communications between the controlling contractor and the entry contractor, and the evaluation of confined spaces that are the same or of a similar configuration. In other words, can multiple houses under construction at a residential community be treated as a single project, or must each house be considered a separate project. In general, it is probably acceptable for a home builder who is acting as the controlling contractor to treat this situation as a single project and perform a single communication to its entry contractor(s), if the potential hazards in the spaces remains the same throughout construction, and the entry contractors remain the same. Likewise, it is also probably reasonable for the entry contractor's competent person to apply their evaluation of a space to the same spaces in the other houses without physically inspecting each one, if there is no reason to believe the conditions or configurations will change. Obviously, a change in entry contractors, a change in the configuration of the space, or the discovery of an unexpected hazard during an entry, would all require the additional communications and evaluations.

8 CONFINED SPACE FATALITIES

During the development of the new OSHA Standards, many asked, "Does the Construction Industry Need These Rules?" In the Federal Register posting that outlines and establishes these new rules, OSHA estimates that the new rules may prevent about 780 serious injuries in the United States each year. Despite this, there were many counter points presented that claimed that the construction industry already had a set of rules they could follow, and others that questioned whether this was even a construction issue. Presumably meaning that confined spaces are primarily found in so called general industry environments and are rare in construction.

Several studies and numerous fatality investigations exist that serve to negate these arguments against needing a new set of mandates.

CASE STUDY NUMBER 1

Scottsdale, AZ on 08/25/2014, AZCentral.com reported that Police believe two workers were overcome by toxic fumes Monday night as they were inside a confined space near a group of restaurants and bars that ring the Promenade shopping center near Frank Lloyd Wright Boulevard and Scottsdale Road in Scottsdale, Arizona. A third man was able to pull himself out of the 15-foot-deep hole before firefighters arrived, according to a Scottsdale police spokesman. That man, an 18-year-old whom authorities would not identify, was "conscious but lethargic" when crews arrived, police said. He could not immediately tell emergency workers on Monday night whether he was pulled or pushed out of the space. Officials said Tuesday that he is expected to recover. The Arizona Division of Occupational Safety and Health (ADOSH) is investigating the incident.

In later testimony given by ADOSH before the Arizona Industrial Commission, it was reported that Camp Industries received a call for emergency maintenance that was needed due to a pump in a sewage lift station that had quit operating. An 18-year old employee was the first to enter the utility vault, and reportedly did so to get the serial or model number from the pump so they would know what they were working with. When the 18-year old employee showed signs of being overcome by a toxic atmosphere, the other two employees of the company that were on-site entered the space to rescue him. These two individuals were the 18-year old boy's father, and the owner of the company. They were able to push the 18-year old out of the space with the help of a bystander who assisted by pulling from above. The two company employees that entered the space to rescue the 18-year old were then overcome by the toxic atmosphere and died. News reports state that first responders measured lethal levels of hydrogen sulfide inside the space, and ADOSH recommended the issuance of citations that identified a failure to properly evaluate the space and identify it as permit-space, a failure to follow a written permit process, and a failure to provide proper training.

The 18-year old later recovered. His father died while rescuing him, along with the owner of the company they both worked for. It was reported that Camp Industries consisted of a total of four employees; the two that died, the one that lived, and the wife of the owner of the company that died in this incident. Testimony before the Industrial Commission of Arizona notes that the employee who first entered the space had received confined space training.

FIGURE 8 – MEMORIAL AT FATALITY SITE

I later had the opportunity to interview several employees, including the management, of another plumbing services company who reported to me that they had regularly serviced this location in the past and had employees inside that very space on several previous occasions. They stated that they were not aware of ever having tested the atmosphere in that space prior to their previous entries and they also stated they had never had any problems in the past with a hazardous atmosphere in the space.

CASE STUDY NUMBER 2

In August 2004 in Kansas City, Missouri, OSHA received notification that one worker had died during the construction of a new sewer manhole. OSHA launched an extensive investigation to determine the cause of the fatality and generated a detailed report and presentation that is available on-line at www.osha.gov. The construction company was laying sewer pipes and installing manholes for a new housing development in what was previously farm land. The land was slightly hilly and the manhole where the fatality occurred was adjacent to a highway on-ramp. There were six employees on-site the day of the fatality.

Per the building code requirements, the company had plugged the two 8-inch sewer lines leading into a manhole in order to perform required vacuum testing on the manhole. The manhole was 4-feet in diameter and approximately 17 feet deep with a typical 2-foot opening at the top. The procedure consisted of pulling a vacuum through the space and maintaining the vacuum at a set pressure for a given amount of time. The company reported that any time the vacuum test failed, the standard procedure was to go inside the manhole and grout the precast concrete manhole segments that make up the manhole riser in order to better seal them. In this case, the initial vacuum test failed and one worker was given a bucket of grout and assigned the task of grouting inside the space while the other five employees worked on other tasks elsewhere on the site. These employees later found the body of the victim at the bottom of the manhole.

After first responders recovered the body, the medical examiner requested air sampling inside the manhole. The fire department's Haz Mat division arrived on site about three hours later to perform air sampling using two direct read multi-gas instruments. The oxygen levels were reported as 16.3% and 17.0%. LEL readings (Lower Explosive Limits of gases) were reported at 0% to 4.5%. There were no detections for hydrogen sulfide or carbon monoxide and no tests were performed at that time for carbon dioxide.

There was conflicting information as to when exactly the test was performed relative to the time of the fatality, but when OSHA arrived on site the day after the incident, they found the grout bucket at the bottom of the hole and the plugs to the sewer line were still in place and under pressure. The manhole cover was replaced, the plug inserts were left in place and the space was left undisturbed for nine days until an OSHA technical crew arrived on site to perform additional air sampling and testing. This was done using both direct read multi-gas instruments and through the bulk sampling and collection of air inside the space and one control sample taken outside the space. The tests produced a variety of results for oxygen that ranged from 12.5 – 18.2%; all below the required 19.5% needed to sustain life. Results for methane gas were 776 – 1372 PPM, which registered on the direct read equipment as LEL readings of 5 – 8%, and carbon dioxide inside the space measured 16,845 – 23,968 PPM, compared to 349 PPM outside the space.

The report notes that the soil in the area is acidic, and theorizes that the acid leaches from the soil and combines with the limestone used in the backfill of the structure to produce carbon dioxide (Note that the concrete structure itself also contains limestone). This carbon dioxide is then drawn into the manhole during the vacuum test, resulting in a toxic atmosphere. Carbon dioxide itself is only hazardous to humans in very high concentrations, but in a confined space it displaces the oxygen in the air and results in an oxygen deficient atmosphere. Prior to this investigation, there were five additional instances of fatalities in the previous four years where people died inside of newly constructed manholes that had not yet been placed into service.

OSHA's investigation revealed that the construction company had no confined space program and did not have any testing or monitoring equipment available at the time the work was being performed. It also includes a note that the contractor later borrowed a direct read multi-gas meter and was inserting it into the space alongside of OSHA's technical investigations team. As they did this, the instrument's alarm would sound to indicate an oxygen deficient atmosphere in the space.

CASE STUDY NUMBER 3

A June 2015 article appearing in Industrial Safety and Hygiene News describes an incident in California that was investigated by the National Institute for Occupational Safety and Health (NIOSH) under its Fatal Accident Circumstances and Epidemiology (FACE) Project. A general engineering contracting firm was hired to replace a valve in a 72-inch underground water line. The water line had separated due to land subsidence and the company had enclosed this portion of the water line in a rectangular concrete vault measuring 12-feet x 15-feet, and approximately 15-feet deep. The concrete service area was covered with a removable wooden roof with a steel access door (3' X 5'). Three vents (approximately 6" X 24") were present on two opposite sides of the service area, above ground level. Inside the service area the 72-inch line was reduced to 54 inches and a valve was installed on the concrete-coated steel water line. One of the tasks remaining was to paint the valve and the steel flanges with an epoxy coating that contained 2-nitropropane and coal tar pitch.

The day prior to the fatality, a foreman and a co-worker worked applying the same epoxy coating to water line support rods in a similar water line service area that was located approximately 300 feet away, but was open at the top. After work that day, the co-worker complained of nausea and a headache, but returned to work the following morning, having apparently recovered. That next morning, they entered the service area, left the hatch open to provide light in the space, and began applying the epoxy coating. During this operation, a third worker and a safety inspector employed by the firm overseeing the project entered the space. By the time the crew took a break to eat lunch, the two workers and the inspector all complained about the "fumes" in the space, but nothing was done to correct the issue and the foreman and the original worker re-entered the space to finish the work.

After their shift was over, both men experienced nausea and headaches on the way home, and the foeman vomited on the way home. Both decided to go to the hospital and were admitted that evening. They were both discharged the following morning, however the foreman was readmitted to the hospital four days later. He lapsed into a coma and died of acute liver failure 10 days after the exposure. The NIOSH report concludes that death was the result of exposure to 2-nitiopropane and coal tar pitch vapors.

The investigation revealed that the epoxy coating was clearly labeled stating it should be used in confined spaces "only with adequate forced air ventilation to prevent dangerous concentrations of vapors which could cause death from breathing." The company had a written policy in its safety plan outlining procedures for work in confined spaces, and there was a blower on site that is used to provide continuous forced air ventilation, however it was not used. In its report, NIOSH states "If these procedures had been followed, the likelihood of this incident occurring would have been reduced, perhaps eliminated."

SUMMARY

Case study Number 1 illustrates a classic example of ever changing conditions, even in spaces that have been entered without incident numerous times in the past. When I interviewed the other plumbing service company I mentioned in the case study, they stated that when news of this fatality broke that afternoon, they absolutely assumed that they were hearing news about their employees since they regularly serviced this location. One of the things they stated several times is that they had never encountered any problems with the air in this space during their past work. This tragic incident served as a wake-up call to them to create a rigid confined space entry program to ensure this never happens to the people they employ.

Case study Number 2 describes a series of conditions and events that combined to create a hazardous atmosphere inside of a newly constructed manhole. The space was clean and the system had not yet been put into service. Despite this, this combination of circumstances led to a condition that created a hazardous atmosphere that resulted in a fatality. This should be an example to the industry that even newly created spaces are not immune to these conditions and should never be automatically considered safe. Structures such as sewers and storm drains are susceptible to infiltration by gases that exist in the surrounding soils as well as gases that can be introduced from above the space, such as exhaust from idling vehicles or equipment. A properly performed test of these spaces is quick, inexpensive, and vital to the safety of the people that will enter to perform work. Unfortunately, the number of construction companies who are currently trained and equipped remains low (in this author's opinion), particularly those companies whose work is mostly limited to new construction.

Case study Number 3 is interesting in that it appears at least some effort was made at some point to acknowledge that the space needed ventilation, and in fact it was reported that mechanical ventilation equipment was actually on-site, but not used. It is also unique from the first two case studies in that the hazard in the space was introduced because of the work itself (applying the coating material). In addition, the instructions for the material being used in the space appears to have clearly identified the hazard, and other employees, including a so-called safety professional, noted a potential hazard but seemingly took no action. This case study also presents an example of a fatality occurring more than a week after the exposure occurred, which needs to be a lesson to all that just because everyone exits the space doesn't mean they are out of danger, particularly when the hazard involves exposure to a toxic atmosphere.

These three incidents cover conditions ranging from new construction to repair and restoration. The fatality in the newly constructed sewer manhole is clearly an example of work defined as "construction" by OSHA. The work being done inside the sewer lift station was considered by ADOSH to be maintenance and they recommended citing the employer under the confined spaces regulations for General Industry, which also allowed them to cite the property owner for failure to perform its duties under that standard. The final incident could probably be characterized either way, as maintenance covered under the General Industry regulations, or as construction which put it, at the time, under a very non-specific set of rules related to confined space entry. This characterization of the work being done has, in the past, created a sort of loop hole in the treatment of some confined spaces, and is one of the things the new rules for confined spaces in construction aims to eliminate.

With the issuance of 29 CFR1926 Subpart AA, Confined Spaces in Construction, entry into confined spaces can no longer be handled in any less formal manner just because one party characterizes the work as construction. In fact, OSHA has stated that in the case of construction work being done at an existing facility that contains permit-required confined spaces, the parties will be considered in compliance if they follow the rules for construction. This is a considerable change in how the industry must approach this work.

Previously, the construction industry was able to approach work inside a confined space by stating that the rules for construction required them only to "train entrants in the hazards…", and any claims that they should follow the more formal entry procedures contained in the General Industry Standards (29 CFR 1910.146) was often met with a response taken directly from those same standards in 29CFR1910.146(a) that states "This section does not apply to agriculture, to construction, or to shipyard employment". While none of this obviated the general duty of an employer to provide a safe and healthy work environment, it did allow a construction employer to enter confined spaces, and even spaces fitting the definition of permit-required confined spaces, under a much less stringent set of rules and procedures. It also meant that compliance enforcement under these conditions was restricted to either enforcement under the failure to train clause contained in §1926.21(b)(6)(i), or citation under the General Duty clause after an incident occurred.

One of the practical implications of this previous situation was that once the work was characterized as construction, the primary entity at risk for enforcement actions by OSHA was the Entry Employer. The rules for multi-employer work sites, also known as the Rules of Construction, do not permit the citation of an owner or general contractor under the General Duty Clause when their employees are not exposed to the hazard, and the only specific requirement for work in confined spaces in construction was the duty for the Entry Employer to provide training. This all changes under the new rules.

9 PREVENTION THROUGH DESIGN

The concept of designing a structure for safety, known as Prevention Through Design (PTD) is one that has received very little attention within the design community in the past. Generally speaking, a designer's goal is to create a finished structure that meets the owners stated needs and complies with all applicable codes and standards. Unfortunately, the codes and standards taken into consideration by the designer are typically limited to building-codes and utilization standards such as accessibility requirements, and few owners delve very deep into taking maintenance into consideration when they communicate their expectations. This is evident in many aspects of design and construction, such as:

- A building's parapet walls rarely meet the height requirement that allows them to be used for fall protection of workers performing maintenance on the roof

- Retaining walls along highways have been constructed without consideration to the fall hazards that will be faced by workers tending to the landscaping in the planters formed at the top of the walls

- Light fixtures are located at the top of multi-story open areas which will require scaffolding in order to safely change the bulbs

Each of these are examples where consideration during design could have resulted in a finished structure that is safer to maintain; extending parapet walls or adding railings could get them to a height that protects workers, and relocating lights so they are more accessible or even replacing them with 50 year LED's can have a positive impact on the safe maintenance of a structure throughout its life.

Designing for safety can also certainly be extended to include considerations for working in confined spaces. Certainly, every project and facility is different in its configuration and needs. Taking this concept into consideration can result in facilities that are not only safer to operate and maintain, but safer to construct as well.

Some examples and considerations related to confined spaces include:

- In residential construction, locating the gas furnace in a closet as opposed to an attic can help rid the attic of a hazard that could necessitate it being classified as permit-required

- Designing equipment that is located within an attic or crawl space with properly located disconnects can assist a service technician by allowing them to shut off and lock out hazardous energy before entering the confined space, and specifying disconnect switches that are designed with integrated lock out features makes it more likely that they will actually be locked out once they are turned off

- Add service lighting to confined spaces where appropriate

- Specify valves and shut off devices that can facilitate isolating and draining lines from outside of a confined space such as a vault that may require entry for maintenance

These are just a few examples that are meant to get a designer thinking about the concept of designing for safety. Another innovative and extremely useful design innovation would be adding an additional access point to a utility vault. This has been done to create a walk-in entry into a utility vault that would typically only have an overhead entry from the street level, with access by a ladder down into the vault. Since the vault was located directly adjacent to the building's basement, a regular access door was added in the basement level to create an alternate access point. In this example, the overhead cover is used to raise and lower equipment into the vault for servicing, but personnel access the space via the walk in door, eliminating any restricted access issues. Designing safety features into a structure can result in more efficient and more affordable maintenance and construction during the life of the structure, but it does require the design team to have an awareness of construction and maintenance safety issues that they typically do not have. Hopefully the increasing widespread use of various levels of integrated project delivery methods that bring construction professionals into the design conversation, will help facilitate more designed for safety features in the future.

10 SUMMARY AND NEXT STEPS

All construction employers should take steps to evaluate their working conditions and evaluate the steps need to take to ensure safe work sites for their employees and compliance with applicable regulations. These steps will vary depending upon the role(s) assumed by the company.

Trade Contractors

Virtually all trade contractors should include the topic of confined spaces in their safety and health management system. At a minimum, all employees need to understand what a confined space is, what the unique hazards are, and they need to understand that they must stay out of them unless specifically authorized and instructed to enter for the purposes of conducting required work. More directly, all employees need to be instructed that they are not to enter any space that is marked as a permit-required confined space unless they are part of an authorized entry team. They should also be directed to question anything that isn't marked, but looks like it could be a confined space. Remember, if it's a confined space and you have to enter to conduct work, the space must be evaluated by a Competent Person for the purpose of designating it a permit-required space or a non-permit space. If it meets the definition of a confined space, it cannot be entered until your Competent Person conducts this evaluation.

Remember that construction sites constantly change and evolve. Consider extending the topic of confined space management to your estimating and contracting activities. If you are planning to perform work in a space that does not meet the definition of a permit-required space, but the project sequencing is rearranged and results in conditions that dictate a Permit entry, this will have a significant impact on your costs and your scheduling.

Be prepared to receive notices from the general contractor about confined spaces, and have a system in place for handling them. If you are notified of permit-required confined spaces on a project and will have people working in the proximity of these spaces, you must ensure they are made aware of this fact and that they know they cannot enter these spaces. Remember that §1926.1202 states...

An employer cannot avoid the duties of the standard merely by refusing to decide whether its employees will enter a permit space, and OSHA will consider the failure to so decide to be an implicit decision to allow employees to enter those spaces if they are working in the proximity of the space.

Finally, make sure your management team understands the meaning of a confined space. If your work results in the construction of a confined space, you are required to notify the Controlling Contractor. Include activities such as setting a new vault or manhole, or enclosing an attic or crawl space. Again, this should be part of a formal management process within your company; be prepared to receive notices of confined spaces on projects and be prepared to provide notification for any spaces you construct.

Controlling Contractors

If you are the general contractor on a construction project, you probably meet the definition of the Controlling Contractor. You might also meet the definition of a Controlling Contractor if you are the property owner and are directly managing the work of multiple contractors. Remember that your employees are not exempt. You need to ensure that they understand how to recognize a confined space and what additional hazards may exist that turn that space into a permit-required space. You will need to decide what your company's policy will be regarding entry into these spaces.

Project management personnel and superintendents are not exempt from compliance; any hazards that exist during entry are there regardless of who enters and regardless of why. Also remember that if entry is required to eliminate or isolate hazards in a space, that entry is required to be done following the procedures for a Permit entry (§1926.1203(g)(2)).

To be clear, there is no provision for an exemption for inspecting work or conditions. Entry is entry, and that means that as the Controlling Contractor you need to make a determination as to whether your employees will enter permit-required spaces and you will need to control and coordinate access to these spaces by third party personnel such as inspectors.

The Controlling Contractor needs to implement a system to facilitate the collection and dissemination of information as explained in Chapter 2. Information needs to be obtained from the Host Employer and transmitted to all subcontractors. Requests need to be sent to all subcontractors asking them to identify any confined spaces they will create during construction, and this accumulated information needs to be transmitted back to all subcontractors on the project. Think of this process as being similar to the collection of Hazard Communication data; i.e. the Safety Data Sheets (SDS's formerly known as MSDS's). Since the expansion of the Haz Comm rules (§1910.1200) in 1987 to the construction industry, it has become standard practice at the beginning of virtually all construction projects for the Controlling Contractor to request SDS's from all subcontractors that will bring hazardous chemicals on to the project site. Those are collected and put into binder on the job site so that the information can be accessed by all trades that might be working in the proximity of chemicals being used on-site by others. As new subcontractors are added to the project, they get a similar request and any SDS's they provide are added to the binder. Confined spaces need to be handled in a similar fashion on every project, although the method of communicating the accumulated information needs to be more affirmative to ensure that affected personnel receive the information.

The Controlling Contractor needs to implement a system under which it ensures that notices are posted and entrance is blocked at all recognized permit-spaces. While some of this actual physical work may be contracted away to an Entry Contractor that will take control of the space, it remains the Controlling Contractors responsibility to bar entrance to any recognized permit-space that exists on the project site unless and until such space(s) is taken over by an Entry Contractor.

The Controlling Contractor must coordinate entry among multiple trades and must ensure that entry into a permit-required space is made only by an authorized Entry Contractor who makes entry under a valid written entry plan and permit that complies with the rules contained in §1926 Subpart AA. During construction, data must be collected from subcontractors relating to all permit-required confined space entries that take place, and that information must be passed on to any subcontractors who will make subsequent entry into the same space. After the project is completed and turned over to the owner (Host Employer), this same information must be transmitted for the owner's inclusion in its confined space program.

Host Employer

The Host Employer has obligations under the OSHA Standards to provide information on known permit-required confined spaces that exist on its property. This includes spaces you know about, and spaces you should know about. The rules for General Industry have always required a property owner to evaluate the workplace and classify any permit-required confined spaces (§1910.146(c)(1)) whether they enter them or not. These rules, in §1910.146(c)(8) have always required information on these spaces be transmitted to any other company, a service company for example, that was hired to enter these permit-required spaces, and this required passing of information now extends to anyone hired to do construction work on the site, whether that work will be conducted in the space, or even just in proximity of the space.

The Host Employer must pass on information not only on the existence of permit-required spaces, but also any information it has been given related to previous entry into these spaces. The intent is to keep the contractors who will work on site as informed as possible. If records exist showing a hazardous atmosphere detected during a previous entry, providing this to the contractors who will perform construction work on site will help them to create a plan for safely entering or working around these spaces.

Entry Contractors

Any company that will direct its employees to enter a permit-required confined space for the purpose of completing work, conducting an inspection, or performing tests is an Entry Contractor subject to all of the rules and provisions contained in §1926 Subpart AA and explained in this book. Remember that Entry is defined in §1926.1204 as meaning the action by which any part of a person passes through an opening into a permit-required confined space. Entry is considered to have occurred as soon as any part of the entrant's body breaks the plane of an opening into the space, whether or not such action is intentional or any work activities are actually performed in the space. This is an important definition and is one of the reasons that general confined space training needs to be provided to all construction employees.

It is important for everyone on a job site to understand that the simple action of opening the hatch to a crawl space that has been designated a permit-required space and sticking your head inside to look at something makes you an Entry Contractor. In this situation, that person could be immediately exposed to the hazards of the space, therefore they must immediately comply with the rules for a Permit entry.

An Entry Contractor is subject to considerable additional training beyond the general hazard awareness training that all personnel should undergo. That training must include the specific hazards of the space they will enter. That training must also be expanded to include the proper use of any required PPE, the specific rescue procedures and any required rescue equipment, and the specific procedures contained in the written Permit that must be followed to render a space acceptable for entry. All this training must be pursuant to an evaluation conducted by its Competent Person and all the procedures to be followed must be documented in writing and available at the permit-space. Entry must be controlled by an Entry Supervisor and overseen by an Entry Attendant whose primary responsibility must be to limit access only to authorized Entrants and then to monitor the safety and health of those Entrants the entire time they are in the space.

FIGURE 9 –EQUIPMENT NEEDED BY AN ENTRY CONTRACTOR

At the conclusion of an entry, the Entry Contractor must document the entry conditions they encountered and must review this information at least annually to ensure that procedures being used are effective and producing the desired results. This information must also be passed to the Controlling Contractor (or directly to the Host Employer in the case of no Controlling Contractor) to facilitate the communication loop described in Chapter 2.

Conclusions

Contractor's who's work requires entry into permit-required confined spaces must read the OSHA regulations contained in §1926 Subpart AA (found in Appendix B) and be familiar with the requirements. This book is meant to provide additional guidance and clarification on the rules contained in those regulations and to provide recommendations for best practices when it comes to managing confined spaces on construction sites. The safest way to manage confined spaces is to find a way to avoid entry into a space that contains recognized hazards. If entry into such a space is required, significant steps must be taken to ensure the safety of the Entrants, including proper evaluation, training, use of PPE and the establishment and adherence to a space specific entry plan developed by a trained and experienced competent person.

Again, the safest way to work inside of a permit-required space is to not work inside of a permit-required confined space. In the construction industry we may have options available to make that happen. The project can sometimes be re-sequenced so that work can be performed inside of a space before the introduction of hazards, or better yet, perform the work before enclosing the space. If those options are not available, try to establish a plan whereby all hazards in the space are eliminated or isolated so the space can be re-classified as a non-permit space. When work must be performed in a permit-required space, preplanning will help to ensure the safest possible entry and will result in the most productive work flow.

APPENDIX A – WORKING WITH ATMOSPHERIC TESTING EQUIPMENT

Many confined spaces share the common concern about the quality of the air they contain. This makes the air quality meter or testing device one of the more common pieces of equipment used by confined space entrants. Despite that, the construction industry continues to see confusion and lack of knowledge in some of the operating aspects of this very important piece of equipment. Specifically, it is still common to see considerable differences of opinion among individuals as to the method and frequency of calibration needed to ensure proper operation of these devices.

Put quite simply, these devices must be tested and calibrated in accordance with the manufacturer's requirements. While that might seem like an oversimplified statement, you might be surprised at how often questions can be answered by simply looking at the instructions. From a regulatory standpoint, the instructions are the defining document: tools and equipment must be used, tested, repaired and calibrated in accordance with the manufacturer's instructions. Many people tend to want to see something from some regulatory body, such as OSHA, that just tells them when the rules require them to calibrate their equipment, but as technology changes and makes rapid advances, it becomes more difficult to issue a single, one-size-fits-all, regulatory document. So, the rules are simply, follow the manufacturer's instructions.

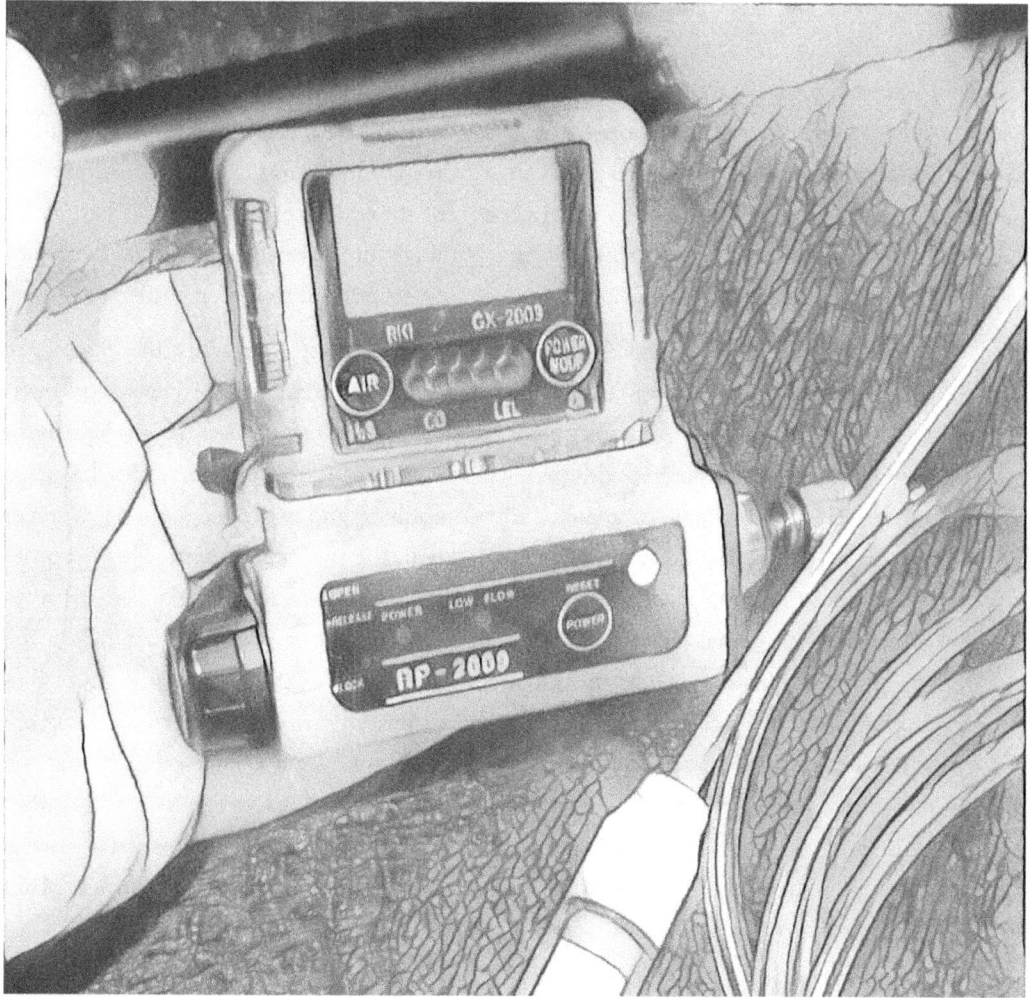

FIUGURE 10 – ATMOSPHERIC MONITOR WITH SAMPLING PUMP ATTACHED

There is a little more guidance to be found with respect to calibrating these atmospheric testing devices, but even it can be confusing. OSHA, NIOSH and other organizations have issued different documents attempting to deal with this issue, but some of them contradict each other, some are unclear, and other have issued updates with information that is confusing to reconcile. Generally, all the guidance documentation set out stating the same thing: equipment must be tested each day before use to ensure that it is in working condition. This includes ensuring that it detects each gas it is configured for, that it detects it and alarms in a timely fashion, and that it correctly reports the detected quantities. The confusion begins with trying to figure how this done and continues into the question of "Do I really need to do this every day?"

Before jumping into that question, it is important to define exactly what equipment is being discussed and exactly what conditions we are concerned about. This Appendix and this discussion will focus on Direct Reading Portable Gas Monitors. Most of these will use diffusion sampling to detect and report levels of different gasses. In the construction industry it is probably most common to purchase and utilize a four (4) gas meter, with a standard configuration to test for Oxygen (O2), Hydrogen Sulfide (H2S), Carbon Monoxide (CO) and Combustible Gasses. The modern version of these devices is very small (handheld) and may contain electronic components that are generally not user serviceable plus a sensor block that may be serviceable. A few items that are very important to understand:

1. Although this is a "standard" configuration because it represents the things we commonly need to test for in confined spaces on construction projects (you must test the oxygen level, you always want to know if explosive gases of any kind have accumulated, carbon monoxide is ever present on construction sites as it is generated by the exhaust of vehicles and equipment, and hydrogen sulfide is a sewer gas) you must match the gasses being tested with the potential hazard. This becomes critical in an existing facility where other gases are known to be produced. If other gasses are a potential hazard, you must test for them to ensure they are below the acceptable standards such as those contained in OSHA §1926-Subpart Z (note that Subpart Z is, in some cases, considered out of date and may not contain a complete and current list of permissible exposure limits for all substances which may be encountered – it may be necessary to consult additional resources such as NIOSH guidelines or to engage the services of a trained industrial hygienist). Special conditions may require additional equipment if your standard equipment isn't configured properly.

2. Testing to ensure exposure levels are below OSHA Permissible Exposure Limits (PEL's) is a minimum regulatory requirement and there may be more recent guidance documents available that recommend more stringent PEL's based on more recent research and data. Sources include the National Institute for Occupational Safety and Health (NIOSH) and the American Industrial Hygiene Association (AIHA). It is advisable to use the most recent recommendations if they are more stringent than the old PEL's published in the regulatory documents.

3. The sensor block in these instruments contain sensors that have a service life – they expire. Know what that service life is and get them replaced as needed. Modern devices contain very small, easily replaceable, self-contained sensor components. They may also contain filters or scrubbers that need to be maintained and replaced. Proper maintenance is just as important as proper calibration.

One of the things that makes answering the question of "How often do I need to calibrate these instruments?" so hard to answer is that the manufacturer's instructions often contain statements such as this…

> *What we generally recommend is that users develop a frequency of calibration that is tailored to their application and usage. Initially, the user may begin by calibrating once per week, and note any changes or adjustments needed to the calibration. If, week after week, there is very little or no adjustment needed, then the calibration frequency can decrease to the point that there will be only a small adjustment needed when calibrating. In general, for most users, this frequency ends up being somewhere between one and three months. For users who do not wish to develop their own frequency, we recommend that they calibrate once a month. For users who "bump test" their instrument prior to each use, the calibration cycle can be extended to 3 to 6 months for instruments that successfully pass the bump gas test.*

This statement, which is the manufacturer's instructions that we are required to follow, contains several terms that need to be defined in order to decipher the requirements. This statement uses the terms bump test and calibration, and in fact much of the available literature splits the term calibration into two terms, calibration check and full calibration. The International Safety Equipment Association (ISEA) defines these three terms as follows:

- **Bump Test** – (Also referred to as a Function Check) A qualitative function check where a challenge gas is passed over the sensor(s) at a concentration and exposure time sufficient to activate all alarm indicators to present at least their lower alarm setting. The purpose of this check is to confirm that gas can get to the sensor(s) and that all the alarms present are functional. This is typically dependent on the response time of the sensor(s) or a minimum level of response achieved, such as 80% of gas concentration applied. Note this check is not intended to provide a measure of calibration accuracy.

- **Calibration Check** - A quantitative test utilizing a known traceable concentration of test gas to demonstrate that the sensor(s) and alarms respond to the gas within manufacturer's acceptable limits. This is typically ±10-20% of the test gas concentration applied unless otherwise specified by the manufacturer, internal company policy, or a regulatory agency.

- **Full Calibration** – The adjustment of the sensor(s) response to match the desired value compared to a known traceable concentration of test gas. This should be done in accordance with the manufacturer's instructions.

If course the idea is simple, you want to make sure the device is working properly since you are risking your life on it. Much of the literature published by regulatory agencies is in agreement that a function check must be performed daily before each use. Again, the idea is to test the device and make sure that gas can reach the sensors, the sensors can detect the gas, and the equipment alarms as desired. Most manufactures will agree to some extent with the ISEA's position statement that the only way to be sure that a device is functioning properly is to perform either a Bump Test or a Calibration Check.

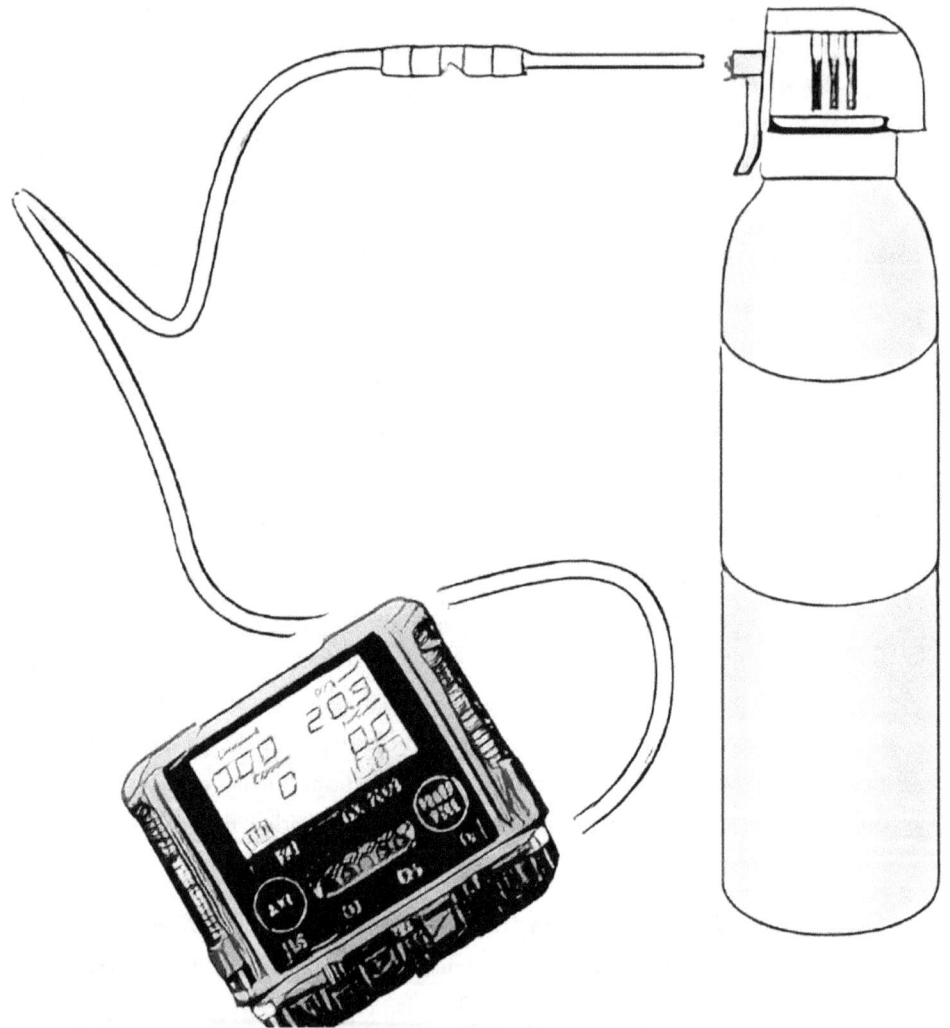

FIGURE 11 – SMALL FUNCTION TEST CYLIDER REQUIRES NO REGULATOR

ISEA's position paper, published March 5, 2010, recommends that a Bump Test or a Calibration Check should be performed before each day's use, and that a Full Calibration should be conducted at regular intervals in accordance with instructions specified by the instrument's manufacturer, internal company policy, or a regulatory agency. Both these tests are easy to perform and they are quick, taking only minutes. They both consist of exposing the sensors to a gas that is expected to activate its alarm. The difference between the two tests are that the Bump Test simply verifies that the sensors are receiving and detecting the gas by setting off the alarm, while the Calibration Check uses a known quantity of gas (usually marked on the gas cylinder) so that you can compare the known value to the actual reading on the instrument.

The key to performing either of these tests lies in the fact that to comply with the intent and actually perform a function check, you must test all gasses. It is common in the construction industry to see a function test performed by simply exposing the equipment to an atmosphere that is known to be safe (fresh air outside) to verify that the reading is about 21% (the theoretical O2 content of fresh air is 19.95%), and then blowing into the sensor which exposes it to a lower volume of oxygen (exhaled breath can be in the range of 11-16% which should trip the low oxygen alarm). While this might serve as an acceptable test of the O2 sensor, if you are Bump Testing the four-gas meter described above, this method does not test the H2S sensor, the CO sensor, or the combustible gas sensor. The only way to truly perform a function test on these sensors is to expose them to a concentration of gas that is above their sensor alarm levels. The simplest, most consistent, and most accurate way of doing this is to use a gas cylinder that contains the target gases. Here, too, technology has produced an extremely simple, efficient and cost effective way to perform this quick function check that is much more accurate and reliable than simply blowing on the sensors. There are several manufacturers that produce a product the size of a small aerosol can, that simply connects to the testing equipment with a plastic tube. It does not require a regulator. All you do is pull the trigger on the can to expose the sensors to all four gases. It is inexpensive and convenient, and one small spray can sized disposable bottle can last for months. This facilitates performing a true function test in a matter of minutes before each use.

To perform a Calibration Check or a Full Calibration, it is necessary to utilize a gas cylinder that contains a known quantity of each type of gas. This will most likely require a more expensive gas cylinder and regulator with the cylinder listing the precise concentration of each type of gas along with the expiration date of the gas. A Calibration Check is performed by reading the meter while the sensors are exposed to the known gas concentrations and then comparing the values. These Calibration Checks should be recorded and documentation will need to be retained to validate and provide proof of calibrated equipment. If an instrument fails a bump test or the anticipated values are not displayed during a Calibration Check, a Full Calibration will need to be performed in accordance with the manufacturer's instructions to adjust the sensor's response to match the known concentration of gas. This procedure may also lead to the conclusion that one or more sensors needs to be replaced.

Again, the desired frequency of the Calibration Checks or Full Calibrations will vary per the manufacturer's instructions. Many will recommend an interval of one month, while others will state that this frequency can be lengthened if daily bump tests are performed. Infrequently used equipment that is only used once every few months may warrant a Calibration Check prior to each use. In any case, establishing a policy that meets the manufacturer's recommendations and suits the service conditions under which the equipment is used is the key to keeping the equipment in a safe working condition. Documenting adherence to your internal policy is the key to demonstrating regulatory compliance. Of course, the more equipment that a company maintains, the more extensive the controls that must be put into place to track calibration and compliance. The introduction of logging features and the use of docking stations can make this process easier. Docking stations can greatly simplify the calibration process and can automate the process of tracking and documentation.

AUTHORS OBSERVATIONS

After many years in the industry I have very rarely seen an actual bump test performed on all of the sensors in the typical four gas meter that I encounter being used by crews in the field. That's not to say that a test wasn't performed in the office before they left, but my observations indicate that companies that own just a few atmospheric monitors typically do not have a program in place to perform and document adequate function testing and calibration. Many of these devices may sit on a shelf for months in between being used and then get sent out without verifying calibration. I have also seen sales representatives give demonstrations in which they describe blowing into the device sensors as being a bump test. Using a device that doesn't work is just as bad as using no device at all and just hoping for the best! Read the recommendations and requirements discussed in this appendix. If you own this equipment, develop a plan and train the people who will use it. If you rent this equipment, insist on obtaining a copy of the last full calibration, and at the very least, bump test it at the facility when you pick it up. Obtain one of the inexpensive aerosol cans of test gas described herein so that you can continue to perform a full function test on all sensors each day of use, and do not use an instrument that fails a bump test on any of its sensors.

This discussion is meant to be a summary of best practices presented in a manner that will assist a company in the development of its own calibration and testing policies that will result in safe and effective equipment that meets regulatory compliance requirements. The OSHA Standards simply state that, "Before an employee enters the space, the internal atmosphere shall be tested, with a calibrated direct-reading instrument." OSHA Compliance

Directive CPL 2.100, "Application of the Permit-Required Confined Spaces (PRCS) Standards, 29 CFR 1910.146" gives guidance to compliance officers as to what is meant by "calibrated":

A testing instrument calibrated in accordance with the manufacturer's recommendations meets this requirement. The best way for an employer to verify calibration is through documentation.

The Canadian Standards Association in CAS C22.2 NO. 152-M1984 (R2001) does include an additional provision for equipment used to detect combustible gases that essentially requires a successful bump test be performed before each day's use, but this only applies to the performance of the combustible gas sensor.

For more information on the calibration of this equipment and discussions on why sensors degrade and fail, search the Industrial Safety and Hygiene News (ISHN) Magazine. An article titled "Gas Detector Calibration" written by Robert Henderson and published January 4, 2012 provides some excellent discussion and information.

MINIMUM SAFE PRACTICES

This appendix was included in this book because checking and monitoring air quality is a need that is common with entry into many types of confined spaces. Technology has made the equipment needed to perform accurate testing readily available and it is fairly easy to learn how to use it correctly. ALWAYS read the manufacturer's instructions and ALWAYS perform a function check (AKA a Bump Test) on each of the sensors before each day's use. Perform a Calibration Check on a monthly basis unless you have justification for lengthening this interval. If the equipment is not used on a regular basis and sits in storage for a few months before it is needed, perform a Calibration Check before putting it back in service. Perform a Full Calibration according to the manufacturer's instructions any time a sensor fails a Bump Test or a Calibration Check, and as recommended by the manufacturer. Get a system in place to make sure the instruments are serviced and maintained according to the manufacturer's instructions. If you own multiple meters that are used regularly, the docking stations that are available today can automate the process of performing calibration checks and documenting the results for each instrument.

APPENDIX B - §1926-SUBPART AA

Authority: 40 U.S.C. 3701 et seq.; 29 U.S.C. 653, 655, 657; Secretary of Labor's Order No. 1-2012 (77 FR 3912); and 29 CFR Part 1911.

§1926.1201 Scope.

(a) This standard sets forth requirements for practices and procedures to protect employees engaged in construction activities at a worksite with one or more confined spaces, subject to the exceptions in paragraph (b) of this section.

Note to paragraph §1926.1201(a). Examples of locations where confined spaces may occur include, but are not limited to, the following: Bins; boilers; pits (such as elevator, escalator, pump, valve or other equipment); manholes (such as sewer, storm drain, electrical, communication, or other utility); tanks (such as fuel, chemical, water, or other liquid, solid or gas); incinerators; scrubbers; concrete pier columns; sewers; transformer vaults; heating, ventilation, and air-conditioning (HVAC) ducts; storm drains; water mains; precast concrete and other pre-formed manhole units; drilled shafts; enclosed beams; vessels; digesters; lift stations; cesspools; silos; air receivers; sludge gates; air preheaters; step up transformers; turbines; chillers; bag houses; and/or mixers/reactors.

(b) Exceptions. This standard does not apply to: (1) Construction work regulated by §1926 subpart P—Excavations. (2) Construction work regulated by §1926 subpart S—Underground Construction, Caissons, Cofferdams and Compressed Air. (3) Construction work regulated by §1926 subpart Y—Diving.

(c) Where this standard applies and there is a provision that addresses a confined space hazard in another applicable OSHA standard, the employer must comply with both that requirement and the applicable provisions of this standard.

§1926.1202 Definitions.

The following terms are defined for the purposes of this subpart only:

Acceptable entry conditions means the conditions that must exist in a permit space, before an employee may enter that space, to ensure that employees can safely enter into, and safely work within, the space.

Attendant means an individual stationed outside one or more permit spaces who assesses the status of authorized entrants and who must perform the duties specified in §1926.1209.

Authorized entrant means an employee who is authorized by the entry supervisor to enter a permit space.

Barrier means a physical obstruction that blocks or limits access.

Blanking or blinding means the absolute closure of a pipe, line, or duct by the fastening of a solid plate (such as a spectacle blind or a skillet blind) that completely covers the bore and that is capable of withstanding the maximum pressure of the pipe, line, or duct with no leakage beyond the plate.

Competent person means one who is capable of identifying existing and predictable hazards in the surroundings or working conditions which are unsanitary, hazardous, or dangerous to employees, and who has the authorization to take prompt corrective measures to eliminate them.

Confined space means a space that:

(1) Is large enough and so configured that an employee can bodily enter it;

(2) Has limited or restricted means for entry and exit; and

(3) Is not designed for continuous employee occupancy.

Control means the action taken to reduce the level of any hazard inside a confined space using engineering methods (for example, by ventilation), and then using these methods to maintain the reduced hazard level. Control also refers to the engineering methods used for this purpose. Personal protective equipment is not a control.

Controlling Contractor is the employer that has overall responsibility for construction at the worksite.

Note. If the controlling contractor owns or manages the property, then it is both a controlling employer and a host employer.

Double block and bleed means the closure of a line, duct, or pipe by closing and locking or tagging two in-line valves and by opening and locking or tagging a drain or vent valve in the line between the two closed valves.

Early-warning system means the method used to alert authorized entrants and attendants that an engulfment hazard may be developing. Examples of early-warning systems include, but are not limited to: alarms activated by remote sensors; and lookouts with equipment for immediately communicating with the authorized entrants and attendants.

Emergency means any occurrence (including any failure of power, hazard control or monitoring equipment) or event, internal or external, to the permit space that could endanger entrants.

Engulfment means the surrounding and effective capture of a person by a liquid or finely divided (flowable) solid substance that can be aspirated to cause death by filling or plugging the respiratory system or that can exert enough force on the body to cause death by strangulation, constriction, crushing, or suffocation.

Entry means the action by which any part of a person passes through an opening into a permit-required confined space. Entry includes ensuing work activities in that space and is considered to have occurred as soon as any part of the entrant's body breaks the plane of an opening into the space, whether or not such action is intentional or any work activities are actually performed in the space.

Entry Employer means any employer who decides that an employee it directs will enter a permit space.

Note. An employer cannot avoid the duties of the standard merely by refusing to decide whether its employees will enter a permit space, and OSHA will consider the failure to so decide to be an implicit decision to allow employees to enter those spaces if they are working in the proximity of the space.

Entry permit (permit) means the written or printed document that is provided by the employer who designated the space a permit space to allow and control entry into a permit space and that contains the information specified in §1926.1206 of this standard.

Entry rescue occurs when a rescue service enters a permit space to rescue one or more employees.

Entry supervisor means the qualified person (such as the employer, foreman, or crew chief) responsible for determining if acceptable entry conditions are present at a permit space where entry is planned, for authorizing entry and overseeing entry operations, and for terminating entry as required by this standard.

Note. An entry supervisor also may serve as an attendant or as an authorized entrant, as long as that person is trained and equipped as required by this standard for each role he or she fills. Also, the duties of entry supervisor may be passed from one individual to another during the course of an entry operation.

Hazard means a physical hazard or hazardous atmosphere. See definitions below.

Hazardous atmosphere means an atmosphere that may expose employees to the risk of death, incapacitation, impairment of ability to self-rescue (that is, escape unaided from a permit space), injury, or acute illness from one or more of the following causes:

(1) Flammable gas, vapor, or mist in excess of 10 percent of its lower flammable limit (LFL);

(2) Airborne combustible dust at a concentration that meets or exceeds its LFL;

Note: This concentration may be approximated as a condition in which the combustible dust obscures vision at a distance of 5 feet (1.52 meters) or less.

(3) Atmospheric oxygen concentration below 19.5 percent or above 23.5 percent;

(4) Atmospheric concentration of any substance for which a dose or a permissible exposure limit is published in Subpart D—Occupational Health and Environmental Control, or in Subpart Z—Toxic and Hazardous Substances, of this part and which could result in employee exposure in excess of its dose or permissible exposure limit;

Note. An atmospheric concentration of any substance that is not capable of causing death, incapacitation, impairment of ability to self-rescue, injury, or acute illness due to its health effects is not covered by this definition.

(5) Any other atmospheric condition that is immediately dangerous to life or health.

Note. For air contaminants for which OSHA has not determined a dose or permissible exposure limit, other sources of information, such as Safety Data Sheets that comply with the Hazard Communication Standard, §1926.59 of this part, published information, and internal documents can provide guidance in establishing acceptable atmospheric conditions.

Host employer means the employer that owns or manages the property where the construction work is taking place.

Note. If the owner of the property on which the construction activity occurs has contracted with an entity for the general management of that property, and has transferred to that entity the information specified in §1203(h)(1), OSHA will treat the contracted management entity as the host employer for as long as that entity manages the property. Otherwise, OSHA will treat the owner of the property as the host employer. In no case will there be more than one host employer.

Hot work means operations capable of providing a source of ignition (for example, riveting, welding, cutting, burning, and heating).

Immediately dangerous to life or health (IDLH) means any condition that would interfere with an individual's ability to escape unaided from a permit space and that poses a threat to life or that would cause irreversible adverse health effects.

Note. Some materials—hydrogen fluoride gas and cadmium vapor, for example—may produce immediate transient effects that, even if severe, may pass without medical attention, but are followed by sudden, possibly fatal collapse 12-72 hours after exposure. The victim "feels normal" after recovery from transient effects until collapse. Such materials in hazardous quantities are considered to be "immediately" dangerous to life or health.

Inerting means displacing the atmosphere in a permit space by a noncombustible gas (such as nitrogen) to such an extent that the resulting atmosphere is noncombustible.

Note. This procedure produces an IDLH oxygen-deficient atmosphere.

Isolate or isolation means the process by which employees in a confined space are completely protected against the release of energy and material into the space, and contact with a physical hazard, by such means as: blanking or blinding; misaligning or removing sections of lines, pipes, or ducts; a double block and bleed system; lockout or tagout of all sources of energy; blocking or disconnecting all mechanical linkages; or placement of barriers to eliminate the potential for employee contact with a physical hazard.

Limited or restricted means for entry or exit means a condition that has a potential to impede an employee's movement into or out of a confined space. Such conditions include, but are not limited to, trip hazards, poor illumination, slippery floors, inclining surfaces and ladders.

Line breaking means the intentional opening of a pipe, line, or duct that is or has been carrying flammable, corrosive, or toxic material, an inert gas, or any fluid at a volume, pressure, or temperature capable of causing injury.

Lockout means the placement of a lockout device on an energy isolating device, in accordance with an established procedure, ensuring that the energy isolating device and the equipment being controlled cannot be operated until the lockout device is removed.

Lower flammable limit or lower explosive limit means the minimum concentration of a substance in air needed for an ignition source to cause a flame or explosion.

Monitor or monitoring means the process used to identify and evaluate the hazards after an authorized entrant enters the space. This is a process of checking for changes that is performed in a periodic or continuous manner after the completion of the initial testing or evaluation of that space.

Non-entry rescue occurs when a rescue service, usually the attendant, retrieves employees in a permit space without entering the permit space.

Non-permit confined space means a confined space that meets the definition of a confined space but does not meet the requirements for a permit-required confined space, as defined in this subpart.

Oxygen deficient atmosphere means an atmosphere containing less than 19.5 percent oxygen by volume.

Oxygen enriched atmosphere means an atmosphere containing more than 23.5 percent oxygen by volume.

Permit-required confined space (permit space) means a confined space that has one or more of the following characteristics: (1) Contains or has a potential to contain a hazardous atmosphere; (2) Contains a material that has the potential for engulfing an entrant; (3) Has an internal configuration such that an entrant could be trapped or asphyxiated by inwardly converging walls or by a floor which slopes downward and tapers to a smaller cross-section; or (4) Contains any other recognized serious safety or health hazard.

Permit-required confined space program (permit space program) means the employer's overall program for controlling, and, where appropriate, for protecting employees from, permit space hazards and for regulating employee entry into permit spaces.

Physical hazard means an existing or potential hazard that can cause death or serious physical damage. Examples include, but are not limited to: explosives (as defined by paragraph (n) of §1926.914, definition of "explosive"); mechanical, electrical, hydraulic and pneumatic energy; radiation; temperature extremes; engulfment; noise; and inwardly converging surfaces. Physical hazard also includes chemicals that can cause death or serious physical damage through skin or eye contact (rather than through inhalation).

Prohibited condition means any condition in a permit space that is not allowed by the permit during the period when entry is authorized. A hazardous atmosphere is a prohibited condition unless the employer can demonstrate that personal protective equipment (PPE) will provide effective protection for each employee in the permit space and provides the appropriate PPE to each employee.

Qualified person means one who, by possession of a recognized degree, certificate, or professional standing, or who by extensive knowledge, training, and experience, has successfully demonstrated his ability to solve or resolve problems relating to the subject matter, the work, or the project.

Representative permit space means a mock-up of a confined space that has entrance openings that are similar to, and is of similar size, configuration, and accessibility to, the permit space that authorized entrants enter.

Rescue means retrieving, and providing medical assistance to, one or more employees who are in a permit space.

Rescue service means the personnel designated to rescue employees from permit spaces.

Retrieval system means the equipment (including a retrieval line, chest or full body harness, wristlets or anklets, if appropriate, and a lifting device or anchor) used for nonentry rescue of persons from permit spaces.

Serious physical damage means an impairment or illness in which a body part is made functionally useless or is substantially reduced in efficiency. Such impairment or illness may be permanent or temporary and includes, but is not limited to, loss of consciousness, disorientation, or other immediate and substantial reduction in mental efficiency. Injuries involving such impairment would usually require treatment by a physician or other licensed health-care professional.

Tagout means:(1) Placement of a tagout device on a circuit or equipment that has been deenergized, in accordance with an established procedure, to indicate that the circuit or equipment being controlled may not be operated until the tagout device is removed; and (2) The employer ensures that (i) tagout provides equivalent protection to lockout, or (ii) that lockout is infeasible and the employer has relieved, disconnected, restrained and otherwise rendered safe stored (residual) energy.

Test or testing means the process by which the hazards that may confront entrants of a permit space are identified and evaluated. Testing includes specifying the tests that are to be performed in the permit space.

Note. Testing enables employers both to devise and implement adequate control measures for the protection of authorized entrants and to determine if acceptable entry conditions are present immediately prior to, and during, entry.

Ventilate or ventilation means controlling a hazardous atmosphere using continuous forced-air mechanical systems that meet the requirements of §1926.57—Ventilation.

§1926.1203 General requirements.

(a)   Before it begins work at a worksite, each employer must ensure that a competent person identifies all confined spaces in which one or more of the employees it directs may work, and identifies each space that is a permit space, through consideration and evaluation of the elements of that space, including testing as necessary.

(b)   If the workplace contains one or more permit spaces, the employer who identifies, or who receives notice of, a permit space must:

(1)   Inform exposed employees by posting danger signs or by any other equally effective means, of the existence and location of, and the danger posed by, each permit space; and

Note to paragraph §1926.1203(b)(1). A sign reading "DANGER -- PERMITREQUIRED CONFINED SPACE, DO NOT ENTER" or using other similar language would satisfy the requirement for a sign.

(2)   Inform, in a timely manner and in a manner other than posting, its employees' authorized representatives and the controlling contractor of the existence and location of, and the danger posed by, each permit space.

(c)   Each employer who identifies, or receives notice of, a permit space and has not authorized employees it directs to work in that space must take effective measures to prevent those employees from entering that permit space, in addition to complying with all other applicable requirements of this standard.

(d)   If any employer decides that employees it directs will enter a permit space, that employer must have a written permit space program that complies with §1926.1204 implemented at the construction site. The written program must be made available prior to and during entry operations for inspection by employees and their authorized representatives.

(e)   An employer may use the alternate procedures specified in paragraph

§1926.1203(e)(2) for entering a permit space only under the conditions set forth in paragraph §1926.1203(e)(1).

(1)   An employer whose employees enter a permit space need not comply with §§1926.1204 through 1206 and §§1926.1208 through 1211, provided that all of the following conditions are met:

(i)   The employer can demonstrate that all physical hazards in the space are eliminated or isolated through engineering controls so that the only hazard posed by the permit space is an actual or potential hazardous atmosphere;

(ii)   The employer can demonstrate that continuous forced air ventilation alone is sufficient to maintain that permit space safe for entry, and that, in the event the ventilation system stops working, entrants can exit the space safely;

(iii) The employer develops monitoring and inspection data that supports the demonstrations required by paragraphs §1926.1203(e)(1)(i) and

§1926.1203(e)(1)(ii);

(iv) If an initial entry of the permit space is necessary to obtain the data required by paragraph §1926.1203(e)(1)(iii), the entry is performed in compliance with §§1926.1204 through 1211 of this standard;

(v)   The determinations and supporting data required by paragraphs §1926.1203(e)(1)(i), (e)(1)(ii), and (e)(1)(iii) are documented by the employer and are made available to each employee who enters the permit space under the terms of paragraph §1926.1203(e) or to that employee's authorized representative; and

(vi) Entry into the permit space under the terms of paragraph §1926.1203(e)(1) is performed in accordance with the requirements of paragraph §1926.1203(e)(2).

Note to paragraph §1926.1203(e)(1). See paragraph §1926.1203(g) for reclassification of a permit space after all hazards within the space have been eliminated.

(2)   The following requirements apply to entry into permit spaces that meet the conditions set forth in paragraph §1926.1203(e)(1):

(i)   Any conditions making it unsafe to remove an entrance cover must be eliminated before the cover is removed.

(ii)   When entrance covers are removed, the opening must be immediately guarded by a railing, temporary cover, or other temporary barrier that will prevent an accidental fall through the opening and that will protect each employee working in the space from foreign objects entering the space.

(iii) Before an employee enters the space, the internal atmosphere must be tested, with a calibrated direct-reading instrument, for oxygen content, for flammable gases and vapors, and for potential toxic air contaminants, in that order. Any employee who enters the space, or that employee's authorized representative, must be provided an opportunity to observe the pre-entry testing required by this paragraph.

(iv) No hazardous atmosphere is permitted within the space whenever any employee is inside the space.

(v) Continuous forced air ventilation must be used, as follows:

(A) An employee must not enter the space until the forced air ventilation has eliminated any hazardous atmosphere;

(B) The forced air ventilation must be so directed as to ventilate the immediate areas where an employee is or will be present within the space and must continue until all employees have left the space;

(C) The air supply for the forced air ventilation must be from a clean source and must not increase the hazards in the space.

(vi) The atmosphere within the space must be continuously monitored unless the entry employer can demonstrate that equipment for continuous monitoring is not commercially available or periodic monitoring is sufficient. If continuous monitoring is used, the employer must ensure that the monitoring equipment has an alarm that will notify all entrants if a specified atmospheric threshold is achieved, or that an employee will check the monitor with sufficient frequency to ensure that entrants have adequate time to escape. If continuous monitoring is not used, periodic monitoring is required. All monitoring must ensure that the continuous forced air ventilation is preventing the accumulation of a hazardous atmosphere. Any employee who enters the space, or that employee's

authorized representative, must be provided with an opportunity to observe the testing required by this paragraph.

(vii) If a hazard is detected during entry:

(A) Each employee must leave the space immediately;

(B) The space must be evaluated to determine how the hazard developed; and

(C) The employer must implement measures to protect employees from the hazard before any subsequent entry takes place.

(viii) The employer must ensure a safe method of entering and exiting the space. If a hoisting system is used, it must be designed and manufactured for personnel hoisting; however, a job-made hoisting system is permissible if it is approved for personnel hoisting by a registered professional engineer, in writing, prior to use.

(ix) The employer must verify that the space is safe for entry and that the preentry measures required by paragraph §1926.1203(e)(2) have been taken, through a written certification that contains the date, the location of the space, and the signature of the person providing the certification. The certification must be made before entry and must be made available to each employee entering the space or to that employee's authorized representative.

(f) When there are changes in the use or configuration of a non-permit confined space that might increase the hazards to entrants, or some indication that the initial evaluation of the space may not have been adequate, each entry employer must have a competent person reevaluate that space and, if necessary, reclassify it as a permit required confined space.

(g) A space classified by an employer as a permit-required confined space may only be reclassified as a non-permit confined space when a competent person determines that all of the applicable requirements in paragraphs §1926.1203(g)(1) through (g)(4) have been met:

(1) If the permit space poses no actual or potential atmospheric hazards and if all hazards within the space are eliminated or isolated without entry into the space (unless the employer can demonstrate that doing so without entry is infeasible), the permit space may be reclassified as a non-permit confined space for as long as the non-atmospheric hazards remain eliminated or isolated;

(2) The entry employer must eliminate or isolate the hazards without entering the space, unless it can demonstrate that this is infeasible. If it is necessary to enter the permit space to eliminate or isolate hazards, such entry must be performed under §§1926.1204 through 1211 of this standard. If testing and inspection during that entry demonstrate that the hazards within the permit space have been eliminated or isolated, the permit space may be reclassified as a non-permit confined space for as long as the hazards remain eliminated or isolated;

Note to paragraph §1926.1203(g)(2). Control of atmospheric hazards through forced air ventilation does not constitute elimination or isolation of the hazards. Paragraph

§1926.1203(e) covers permit space entry where the employer can demonstrate that forced air ventilation alone will control all hazards in the space.

(3) The entry employer must document the basis for determining that all hazards in a permit space have been eliminated or isolated, through a certification that contains the date, the location of the space, and the signature of the person making the determination. The certification must be made available to each employee entering the space or to that employee's authorized representative; and

(4) If hazards arise within a permit space that has been reclassified as a non-permit space under paragraph §1926.1203(g), each employee in the space must exit the space. The entry employer must then reevaluate the space and reclassify it as a permit space as appropriate in accordance with all other applicable provisions of this standard.

(h) Permit Space Entry Communication and Coordination:

(1) Before entry operations begin, the host employer must provide the following information, if it has it, to the controlling contractor:

(i) The location of each known permit space;

(ii) The hazards or potential hazards in each space or the reason it is a permit space; and

(iii) Any precautions that the host employer or any previous controlling contractor or entry employer implemented for the protection of employees in the permit space.

(2) Before entry operations begin, the controlling contractor must:

(i) Obtain the host employer's information about the permit space hazards and previous entry operations; and

(ii) Provide the following information to each entity entering a permit space and any other entity at the worksite whose activities could foreseeably result in a hazard in the permit space:

(A) The information received from the host employer;

(B) Any additional information the controlling contractor has about the subjects listed in paragraph (h)(1) of this section; and

(C) The precautions that the host employer, controlling contractor, or other entry employers implemented for the protection of employees in the permit spaces.

(3) Before entry operations begin, each entry employer must:

(i) Obtain all of the controlling contractor's information regarding permit space hazards and entry operations; and

(ii) Inform the controlling contractor of the permit space program that the entry employer will follow, including any hazards likely to be confronted or created in each permit space.

(4) The controlling contractor and entry employer(s) must coordinate entry operations when:

(i) More than one entity performs permit space entry at the same time; or

(ii) Permit space entry is performed at the same time that any activities that could foreseeably result in a hazard in the permit space are performed.

(5) After entry operations:

(i) The controlling contractor must debrief each entity that entered a permit space regarding the permit space program followed and any hazards confronted or created in the permit space(s) during entry operations;

(ii) The entry employer must inform the controlling contractor in a timely manner of the permit space program followed and of any hazards confronted or created in the permit space(s) during entry operations; and

(iii) The controlling contractor must apprise the host employer of the information exchanged with the entry entities pursuant to this subparagraph.

Note to paragraph §1926.1203(h). Unless a host employer or controlling contractor has or will have employees in a confined space, it is not required to enter any confined space to collect the information specified in this paragraph (h).

(iv) If there is no controlling contractor present at the worksite, the requirements for, and role of, controlling contactors in §1926.1203 must be fulfilled by the host employer or other employer who arranges to have employees of another employer perform work that involves permit space entry.

§1926.1204 Permit-Required Confined Space Program.

Each entry employer must:

(a) Implement the measures necessary to prevent unauthorized entry;

(b) Identify and evaluate the hazards of permit spaces before employees enter them;

(c) Develop and implement the means, procedures, and practices necessary for safe permit space entry operations, including, but not limited to, the following:

(1) Specifying acceptable entry conditions;

(2) Providing each authorized entrant or that employee's authorized representative with the opportunity to observe any monitoring or testing of permit spaces;

(3) Isolating the permit space and physical hazard(s) within the space;

(4) Purging, inerting, flushing, or ventilating the permit space as necessary to eliminate or control atmospheric hazards;

Note to paragraph §1204(c)(4). When an employer is unable to reduce the atmosphere below 10 percent LFL, the employer may only enter if the employer

inerts the space so as to render the entire atmosphere in the space noncombustible, and the employees use PPE to address any other atmospheric hazards (such as oxygen deficiency), and the employer eliminates or isolates all physical hazards in the space.

(5) Determining that, in the event the ventilation system stops working, the monitoring procedures will detect an increase in atmospheric hazard levels in sufficient time for the entrants to safely exit the permit space;

(6) Providing pedestrian, vehicle, or other barriers as necessary to protect entrants from external hazards;

(7) Verifying that conditions in the permit space are acceptable for entry throughout the duration of an authorized entry, and ensuring that employees are not allowed to enter into,

or remain in, a permit space with a hazardous atmosphere unless the employer can demonstrate that personal protective equipment (PPE) will provide effective protection for each employee in the permit space and provides the appropriate PPE to each employee; and

(8) Eliminating any conditions (for example, high pressure) that could make it unsafe to remove an entrance cover.

(d) Provide the following equipment (specified in paragraphs §1926.1204(d)(1) through (d)(9)) at no cost to each employee, maintain that equipment properly, and ensure that each employee uses that equipment properly:

(1) Testing and monitoring equipment needed to comply with paragraph §1926.1204(e);

(2) Ventilating equipment needed to obtain acceptable entry conditions;

(3) Communications equipment necessary for compliance with paragraphs §1926.1208(c) and §1926.1209(e), including any necessary electronic communication equipment for attendants assessing entrants' status in multiple spaces;

(4) Personal protective equipment insofar as feasible engineering and work-practice controls do not adequately protect employees;

Note to paragraph §1926.1204(d)(4). The requirements of subpart E of this part and other PPE requirements continue to apply to the use of PPE in a permit space. For example, if employees use respirators, then the respirator requirements in §1926.103 (Respiratory protection) must be met.

(5) Lighting equipment that meets the minimum illumination requirements in §1926.56, that is approved for the ignitable or combustible properties of the specific gas, vapor, dust, or fiber that will be present, and that is sufficient to enable employees to see well enough to work safely and to exit the space quickly in an emergency;

(6) Barriers and shields as required by paragraph §1926.1204(c)(4);

(7) Equipment, such as ladders, needed for safe ingress and egress by authorized entrants;

(8) Rescue and emergency equipment needed to comply with paragraph §1926.1204(i), except to the extent that the equipment is provided by rescue services; and

(9)   Any other equipment necessary for safe entry into, safe exit from, and rescue from, permit spaces.

(e)   Evaluate permit space conditions in accordance with the following paragraphs (e)(1) through (6) of this section when entry operations are conducted:

(1)   Test conditions in the permit space to determine if acceptable entry conditions exist before changes to the space's natural ventilation are made, and before entry is authorized to begin, except that, if an employer demonstrates that isolation of the space is infeasible because the space is large or is part of a continuous system (such as a sewer), the employer must:

(i)   Perform pre-entry testing to the extent feasible before entry is authorized; and,

(ii)   If entry is authorized, continuously monitor entry conditions in the areas where authorized entrants are working, except that employers may use periodic monitoring in accordance with paragraph §1926.1204(e)(2) for monitoring an atmospheric hazard if they can demonstrate that equipment for continuously monitoring that hazard is not commercially available;

(iii)   Provide an early-warning system that continuously monitors for nonisolated engulfment hazards. The system must alert authorized entrants and attendants in sufficient time for the authorized entrants to safely exit the space.

(2)   Continuously monitor atmospheric hazards unless the employer can demonstrate that the equipment for continuously monitoring a hazard is not commercially available or that periodic monitoring is of sufficient frequency to ensure that the atmospheric hazard is being controlled at safe levels. If continuous monitoring is not used, periodic monitoring is required with

sufficient frequency to ensure that acceptable entry conditions are being maintained during the course of entry operations;

(3)   When testing for atmospheric hazards, test first for oxygen, then for combustible gases and vapors, and then for toxic gases and vapors;

(4)   Provide each authorized entrant or that employee's authorized representative an opportunity to observe the pre-entry and any subsequent testing or monitoring of permit spaces;

(5) Reevaluate the permit space in the presence of any authorized entrant or that employee's authorized representative who requests that the employer conduct such reevaluation because there is some indication that the evaluation of that space may not have been adequate; and

(6) Immediately provide each authorized entrant or that employee's authorized representative with the results of any testing conducted in accordance with §1926.1204 of this standard.

(f) Provide at least one attendant outside the permit space into which entry is authorized for the duration of entry operations;

(1) Attendants may be assigned to more than one permit space provided the duties described in §1926.1209 of this standard can be effectively performed for each permit space.

(2) Attendants may be stationed at any location outside the permit space as long as the duties described in §1926.1209 of this standard can be effectively performed for each permit space to which the attendant is assigned.

(g) If multiple spaces are to be assigned to a single attendant, include in the permit program the means and procedures to enable the attendant to respond to an emergency affecting one or more of those permit spaces without distraction from the attendant's responsibilities under §1926.1209 of this standard;

(h) Designate each person who is to have an active role (as, for example, authorized entrants, attendants, entry supervisors, or persons who test or monitor the atmosphere in a permit space) in entry operations, identify the duties of each such employee, and provide each such employee with the training required by

§1926.1207 of this standard;

(i) Develop and implement procedures for summoning rescue and emergency services

(including procedures for summoning emergency assistance in the event of a failed non-entry rescue), for rescuing entrants from permit spaces, for providing necessary emergency services to rescued employees, and for preventing unauthorized personnel from attempting a rescue;

(j) Develop and implement a system for the preparation, issuance, use, and cancellation of entry permits as required by this standard, including the safe termination of entry operations under both planned and emergency conditions;

(k)   Develop and implement procedures to coordinate entry operations, in consultation with the controlling contractor, when employees of more than one employer are working simultaneously in a permit space or elsewhere on the worksite where their activities could, either alone or in conjunction with the activities within a permit space, foreseeably result in a hazard within the confined space, so that employees of one employer do not endanger the employees of any other employer;

(l)   Develop and implement procedures (such as closing off a permit space and canceling the permit) necessary for concluding the entry after entry operations have been completed;

(m)   Review entry operations when the measures taken under the permit space program may not protect employees and revise the program to correct deficiencies found to exist before subsequent entries are authorized; and

Note to paragraph §1926.1204(m). Examples of circumstances requiring the review of the permit space program include, but are not limited to: any unauthorized entry of a permit space, the detection of a permit space hazard not covered by the permit, the detection of a condition prohibited by the permit, the occurrence of an injury or near-miss during entry, a change in the use or configuration of a permit space, and employee complaints about the effectiveness of the program.

(n)   Review the permit space program, using the canceled permits retained under paragraph §1926.1205(f), within 1 year after each entry and revise the program as necessary to ensure that employees participating in entry operations are protected from permit space hazards.

Note to paragraph §1926.1204(n). Employers may perform a single annual review covering all entries performed during a 12-month period. If no entry is performed during a 12-month period, no review is necessary.

§1926.1205 Permitting Process.

(a)   Before entry is authorized, each entry employer must document the completion of measures required by paragraph §1926.1204(c) of this standard by preparing an entry permit.

(b)   Before entry begins, the entry supervisor identified on the permit must sign the entry permit to authorize entry.

(c)   The completed permit must be made available at the time of entry to all authorized entrants or their authorized representatives, by posting it at the entry portal or by any other equally effective means, so that the entrants can confirm that pre-entry preparations have been completed.

(d)   The duration of the permit may not exceed the time required to complete the assigned task or job identified on the permit in accordance with paragraph §1926.1206(b) of this standard.

(e)   The entry supervisor must terminate entry and take the following action when any of the following apply:

(1)   Cancel the entry permit when the entry operations covered by the entry permit have been completed; or

(2)   Suspend or cancel the entry permit and fully reassess the space before allowing reentry when a condition that is not allowed under the entry permit arises in or near the permit space and that condition is temporary in nature and does not change the configuration of the space or create any new hazards within it; and

(3)   Cancel the entry permit when a condition that is not allowed under the entry permit arises in or near the permit space and that condition is not covered by subparagraph (e)(2) of this section.

(f)   The entry employer must retain each canceled entry permit for at least 1 year to facilitate the review of the permit-required confined space program required by paragraph §1926.1204(n) of this standard. Any problems encountered during an entry operation must be noted on the pertinent permit so that appropriate revisions to the permit space program can be made.

§1926.1206 Entry permit.

The entry permit that documents compliance with this section and authorizes entry to a permit space must identify:

(a)   The permit space to be entered;

(b)   The purpose of the entry;

(c)   The date and the authorized duration of the entry permit;

(d) The authorized entrants within the permit space, by name or by such other means (for example, through the use of rosters or tracking systems) as will enable the attendant to determine quickly and accurately, for the duration of the permit, which authorized entrants are inside the permit space;

Note to paragraph §1926.1206(d). This requirement may be met by inserting a reference on the entry permit as to the means used, such as a roster or tracking system, to keep track of the authorized entrants within the permit space.

(e) Means of detecting an increase in atmospheric hazard levels in the event the ventilation system stops working;

(f) Each person, by name, currently serving as an attendant;

(g) The individual, by name, currently serving as entry supervisor, and the signature or initials of each entry supervisor who authorizes entry;

(h) The hazards of the permit space to be entered;

(i) The measures used to isolate the permit space and to eliminate or control permit space hazards before entry;

Note to paragraph §1926.1206(i). Those measures can include, but are not limited to, the lockout or tagging of equipment and procedures for purging, inerting, ventilating, and flushing permit spaces.

(j) The acceptable entry conditions;

(k) The results of tests and monitoring performed under paragraph §1926.1204(e) of this standard, accompanied by the names or initials of the testers and by an indication of when the tests were performed;

(l) The rescue and emergency services that can be summoned and the means (such as the equipment to use and the numbers to call) for summoning those services;

(m) The communication procedures used by authorized entrants and attendants to maintain contact during the entry;

(n)   Equipment, such as personal protective equipment, testing equipment, communications equipment, alarm systems, and rescue equipment, to be provided for compliance with this standard;

(o)   Any other information necessary, given the circumstances of the particular confined space, to ensure employee safety; and

(p)   Any additional permits, such as for hot work, that have been issued to authorize work in the permit space.

§1926.1207 Training.

(a)   The employer must provide training to each employee whose work is regulated by this standard, at no cost to the employee, and ensure that the employee possesses the understanding, knowledge, and skills necessary for the safe performance of the duties assigned under this standard. This training must result in an understanding of the hazards in the permit space and the methods used to isolate, control or in other ways protect employees from these hazards, and for those employees not authorized to perform entry rescues, in the dangers of attempting such rescues.

(b)   Training required by this section must be provided to each affected employee:

(1)   In both a language and vocabulary that the employee can understand;

(2)   Before the employee is first assigned duties under this standard;

(3)   Before there is a change in assigned duties;

(4)   Whenever there is a change in permit space entry operations that presents a hazard about which an employee has not previously been trained; and

(5)   Whenever there is any evidence of a deviation from the permit space entry procedures required by paragraph §1926.1204(c) of this standard or there are inadequacies in the employee's knowledge or use of these procedures.

(c)   The training must establish employee proficiency in the duties required by this standard and must introduce new or revised procedures, as necessary, for compliance with this standard.

(d) The employer must maintain training records to show that the training required by paragraphs §1926.1207(a) through (c) of this standard has been accomplished. The training records must contain each employee's name, the name of the trainers, and the dates of training. The documentation must be available for inspection by

employees and their authorized representatives, for the period of time the employee is employed by that employer.

§1926.1208 Duties of authorized entrants.

The entry employer must ensure that all authorized entrants:

(a) Are familiar with and understand the hazards that may be faced during entry, including information on the mode, signs or symptoms, and consequences of the exposure;

(b) Properly use equipment as required by paragraph §1926.1204(d) of this standard;

(c) Communicate with the attendant as necessary to enable the attendant to assess entrant status and to enable the attendant to alert entrants of the need to evacuate the space as required by paragraph §1926.1209(f) of this standard;

(d) Alert the attendant whenever:

(1) There is any warning sign or symptom of exposure to a dangerous situation; or

(2) The entrant detects a prohibited condition; and

(e) Exit from the permit space as quickly as possible whenever:

(1) An order to evacuate is given by the attendant or the entry supervisor;

(2) There is any warning sign or symptom of exposure to a dangerous situation;

(3) The entrant detects a prohibited condition; or

(4) An evacuation alarm is activated.

§1926.1209 Duties of attendants.

The entry employer must ensure that each attendant:

(a)   Is familiar with and understands the hazards that may be faced during entry, including information on the mode, signs or symptoms, and consequences of the exposure;

(b)   Is aware of possible behavioral effects of hazard exposure in authorized entrants;

(c)   Continuously maintains an accurate count of authorized entrants in the permit space and ensures that the means used to identify authorized entrants under paragraph 1926.1206(d) of this standard accurately identifies who is in the permit space;

(d)   Remains outside the permit space during entry operations until relieved by another attendant;

Note to paragraph §1926.1209(d). Once an attendant has been relieved by another attendant, the relieved attendant may enter a permit space to attempt a rescue when the employer's permit space program allows attendant entry for rescue and the attendant has been trained and equipped for rescue operations as required by paragraph §1926.1211(a).

(e)   Communicates with authorized entrants as necessary to assess entrant status and to alert entrants of the need to evacuate the space under paragraph §1926.1208(e);

(f)   Assesses activities and conditions inside and outside the space to determine if it is safe for entrants to remain in the space and orders the authorized entrants to evacuate the permit space immediately under any of the following conditions:

(1)   If there is a prohibited condition;

(2)   If the behavioral effects of hazard exposure are apparent in an authorized entrant;

(3)   If there is a situation outside the space that could endanger the authorized entrants; or

(4)   If the attendant cannot effectively and safely perform all the duties required under §1926.1209 of this standard;

(g)   Summons rescue and other emergency services as soon as the attendant determines that authorized entrants may need assistance to escape from permit space hazards;

(h)   Takes the following actions when unauthorized persons approach or enter a permit space while entry is underway:

(1)   Warns the unauthorized persons that they must stay away from the permit space;

(2)   Advises the unauthorized persons that they must exit immediately if they have entered the permit space; and

(3)   Informs the authorized entrants and the entry supervisor if unauthorized persons have entered the permit space;

(i)   Performs non-entry rescues as specified by the employer's rescue procedure; and

(j)   Performs no duties that might interfere with the attendant's primary duty to assess and protect the authorized entrants.

§1926.1210 Duties of entry supervisors.

The entry employer must ensure that each entry supervisor:

(a)   Is familiar with and understands the hazards that may be faced during entry, including information on the mode, signs or symptoms, and consequences of the exposure;

(b)   Verifies, by checking that the appropriate entries have been made on the permit, that all tests specified by the permit have been conducted and that all procedures and equipment specified by the permit are in place before endorsing the permit and allowing entry to begin;

(c)   Terminates the entry and cancels or suspends the permit as required by paragraph 1926.1205(e) of this standard;

(d)   Verifies that rescue services are available and that the means for summoning them are operable, and that the employer will be notified as soon as the services become unavailable;

(e)   Removes unauthorized individuals who enter or who attempt to enter the permit space during entry operations; and

(f)   Determines, whenever responsibility for a permit space entry operation is transferred, and at intervals dictated by the hazards and operations performed within the space, that entry operations remain consistent with terms of the entry permit and that acceptable entry conditions are maintained.

§1926.1211 Rescue and emergency services.

(a)   An employer who designates rescue and emergency services, pursuant to paragraph §1926.1204(i) of this standard, must:

(1)   Evaluate a prospective rescuer's ability to respond to a rescue summons in a timely manner, considering the hazard(s) identified;

Note to paragraph §1926.1211(a)(1). What will be considered timely will vary according to the specific hazards involved in each entry. For example, §1926.103—Respiratory Protection requires that employers provide a standby person or persons capable of immediate action to rescue employee(s) wearing respiratory protection while in work areas defined as IDLH atmospheres.

(2)   Evaluate a prospective rescue service's ability, in terms of proficiency with rescue-related tasks and equipment, to function appropriately while rescuing entrants from the particular permit space or types of permit spaces identified;

(3)   Select a rescue team or service from those evaluated that:

(i)   Has the capability to reach the victim(s) within a time frame that is appropriate for the permit space hazard(s) identified;

(ii)   Is equipped for, and proficient in, performing the needed rescue services;

(iii)   Agrees to notify the employer immediately in the event that the rescue service becomes unavailable;

(4)   Inform each rescue team or service of the hazards they may confront when called on to perform rescue at the site; and

(5)   Provide the rescue team or service selected with access to all permit spaces from which rescue may be necessary so that the rescue team or service can develop appropriate rescue plans and practice rescue operations.

(b) An employer whose employees have been designated to provide permit space rescue and/or emergency services must take the following measures and provide all equipment and training at no cost to those employees:

(1) Provide each affected employee with the personal protective equipment (PPE) needed to conduct permit space rescues safely and train each affected employee so the employee is proficient in the use of that PPE;

(2) Train each affected employee to perform assigned rescue duties. The employer must ensure that such employees successfully complete the training required and establish proficiency as authorized entrants, as provided by §§1926.1207 and 1926.1208 of this standard;

(3) Train each affected employee in basic first aid and cardiopulmonary resuscitation (CPR). The employer must ensure that at least one member of the rescue team or service holding a current certification in basic first aid and CPR is available; and

(4) Ensure that affected employees practice making permit space rescues before attempting an actual rescue, and at least once every 12 months, by means of simulated rescue operations in which they remove dummies, manikins, or actual persons from the actual permit spaces or from representative permit spaces, except practice rescue is not required where the affected employees properly performed a rescue operation during the last 12 months in the same permit space the authorized entrant will enter, or in a similar permit space. Representative permit spaces must, with respect to opening size, configuration, and accessibility, simulate the types of permit spaces from which rescue is to be performed.

(c) Non-entry rescue is required unless the retrieval equipment would increase the overall risk of entry or would not contribute to the rescue of the entrant. The employer must designate an entry rescue service whenever non-entry rescue is not selected. Whenever non-entry rescue is selected, the entry employer must ensure that retrieval systems or methods are used whenever an authorized entrant enters a permit space, and must confirm, prior to entry, that emergency assistance would be available in the event that non-entry rescue fails. Retrieval systems must meet the following requirements:

(1) Each authorized entrant must use a chest or full body harness, with a retrieval line attached at the center of the entrant's back near shoulder level, above the entrant's head, or at another point which the employer can establish presents a profile small enough for the successful removal of the entrant. Wristlets or anklets may be used in lieu of the chest or full body harness if the employer can demonstrate that the use of a chest or full body harness is infeasible or creates a greater hazard and that the use of wristlets or anklets is the safest and most effective alternative.

(2)   The other end of the retrieval line must be attached to a mechanical device or fixed point outside the permit space in such a manner that rescue can begin as soon as the rescuer becomes aware that rescue is necessary. A mechanical device must be available to retrieve personnel from vertical type permit spaces more than 5 feet (1.52 meters) deep.

(3)   Equipment that is unsuitable for retrieval must not be used, including, but not limited to, retrieval lines that have a reasonable probability of becoming entangled with the retrieval lines used by other authorized entrants, or retrieval lines that will not work due to the internal configuration of the permit space.

(d)   If an injured entrant is exposed to a substance for which a Safety Data Sheet (SDS) or other similar written information is required to be kept at the worksite, that SDS or written information must be made available to the medical facility treating the exposed entrant.

§1926.1212 Employee participation.

(a)   Employers must consult with affected employees and their authorized representatives on the development and implementation of all aspects of the permit space program required by §1926.1203 of this standard.

(b)   Employers must make available to each affected employee and his/her authorized representatives all information required to be developed by this standard.

§1926.1213 Provision of documents to Secretary.

For each document required to be retained in this standard, the retaining employer must make the document available on request to the Secretary of Labor or the Secretary's designee.

ABOUT THE AUTHOR

Jim Rogers is an experienced construction industry trainer and educator that stresses the importance of integrating safety, quality and productivity into all operations. Jim's industry experience includes many years in various roles within the construction industry as well as extensive work with industry trade associations creating training and certification programs. Jim was a faculty member at Arizona State University (ASU) where he taught undergraduate and graduate courses in construction management and safety. While working at ASU he created and ran the University's OSHA Training Institute Education Center in cooperation with the US Department of Labor. He now works with various groups that include:

- LinkedIn Learning where he develops and produces on-line continuing education courses for the Architecture, Engineering and Construction (AEC) industries

- ProCore (construction management software) where he creates support and educational materials related to safety and productivity

- Various trade associations delivering courses in safety and health and general construction management

Jim continues to work with industry to deliver courses, books and articles to advance the knowledge of the construction management professional.

www.ingramcontent.com/pod-product-compliance
Lightning Source LLC
Chambersburg PA
CBHW080703190526
45169CB00006B/2226